Excursions in Astronomical Optics

Springer
*New York*
*Berlin*
*Heidelberg*
*Barcelona*
*Budapest*
*Hong Kong*
*London*
*Milan*
*Paris*
*Santa Clara*
*Singapore*
*Tokyo*

Lawrence Mertz

# Excursions in Astronomical Optics

With 107 Illustrations

Springer

Lawrence Mertz
287 Fairfield Court
Palo Alto, CA 94306
USA

Library of Congress-in-Publication Data
Mertz, Lawrence.
Excursions in astronomical optics / Lawrence N. Mertz.
p. cm.
Includes bibliographical references and index.
ISBN 0-387-94664-0 (hardcover: alk. paper)
1. Astronomical instruments. 2. Optics. 3. Telescopes.
4. Interferometry. I. Title.
QB86.M47 1996
522'.2–dc20 95-51497

Printed on acid-free paper.



Production managed by Frank Ganz; manufacturing supervised by Joe Quatela.
Camera-ready copy prepared from the author 's TEX files.
Printed and bound by Edwards Brothers, Inc., Ann Arbor, MI.
Printed in the United States of America.

9 8 7 6 5 4 3 2 1

ISBN 0-387-94664-0 Springer-Verlag New York Berlin Heidelberg SPIN 10526943

In memory of my son,

**Thomas Mertz**
1965–1980

# Preface

For every astronomical topic that I have approached there has turned out to be a broader realm of possibilities than is commonly accepted or acknowledged. The "excursions" of this book are the examples. They mostly depart from the mainstream of conventional wisdom to offer a wider perspective with opportunities for further research. While my intent is to supplement that mainstream, the effect may appear to dismiss rather than to reconsider accepted tenets. Ample praise and credit for those accomplishments are already available in textbooks. Readers may very well disagree with some of the notions presented in these excursions, but I hope that they will pause long enough to evaluate the scientific basis for any disagreement.

For the most part, these excursions remain incomplete and unfulfilled, yet they contain many ideas that are not available elsewhere. Whether these ideas are perceived as a collection of unproven claims or as a storehouse of fresh opportunities will depend entirely on the attitude of the reader.

The excursions do cover a rather wide span of disciplines, and that may lead to an unfocused overall impression. My hope is thereby to attract a broader audience than that of a single discipline, and to expose them to neighboring disciplines. The excursions all do have the common thread of optical science related to astronomy. The intended audience is workers at the graduate, engineering, or research level engaged in one or more of the various fields. What I have tried to do in each case is to provide the reader with sufficient information to accomplish the unusual projects that are described. For example, the programs given in the Appendices will let the reader make for himself the content of illustrations found in the first two chapters, and then to make his own variations.

Chapter 4 is admittedly rather bland in that the directions have been largely overtaken by more recent advances in image processing, adaptive optics, and laser guide stars. The only reason for retaining it is that there still may be a few items of interest.

Chapters 6 and 7 may seem incongruous with the other excursions, intruding on theoretical turf. They are simply interpretations of the observations from the perspective of an experimentalist in optical science. That extra perspective offers fresh alternatives where theorists may have been too quick in jumping to conclusions with their explanations of pulsars and quasars. At any rate, I hope that these chapters give some grist for thought.

This book is a sequel to my prior book, *Transformations in Optics* [Wiley, 1965], and chronicles my research activities since then. Those activities delimit the scope of the topics. They have been fascinating and stimulating for me. Some of the references mentioned in the Bibliographies are not explicit in the text, but they nevertheless played a role in influencing the text. Also, the Bibliographies are not as thorough nor as up to date as they ought to be because my access to libraries is no longer as convenient as it used to be. As with most authors, there is excess reference to my own contributions simply because of familiarity. I have benefited from conversations with many friends and colleagues over these years. The most longstanding have concerned interferometry with Gerry Wyntjes, formerly of Block Associates and then of Optra. I am grateful to him and to the many others who have encouraged and influenced this work.

August 1994 *L. Mertz*

# Contents

# 1

# Optical Telescopes

## Introduction

Ever since Galileo first aimed a telescope upward there has been a continual quest to examine the sky in ever-increasing detail. The optical quality of objective lenses largely restricted the performance of those early telescopes. In the latter half of the 17th century, the Huygens brothers recognized that long focal lengths were less susceptible to aberrations. That recognition inspired telescopes with 30- to 60-meter focal lengths, but such extremes proved so unwieldy that more modest sizes in the range 6- to 10-meter focal lengths were more effective in practice. Quite later, after mastery of the worst aberrations, diffraction was noticed as an aperture dependent restriction on resolving power. The subsequent advent of photography further complicated the interplay of focal length and aperture as they relate to telescope performance, because photographic emulsions, with their thresholds and integrating character, behave quite differently than visual observation. Much more recently, the interplay has been modified for infrared and charge-coupled-device (CCD) sensors. In any case, it is only over the last century that the two most notorious factors that limit telescope performance have become conspicuous. The first is atmospheric turbulence, and the second is the quantized nature of light.

The actual meaning of performance depends somewhat on the task to be accomplished, whether it be to examine a planet in utmost detail or to study ever-fainter objects far beyond the solar system. The space era now has accomplished the first task so overwhelmingly that the latter task now predominates for ground-based telescopes. Nevertheless, there still exists a critical dependence on resolving power.

Consider the problem of detecting a faint star. Obviously, the more stellar photons gathered, the better. Therefore, the telescope should have as large a collecting area as possible, and the observation should last as long as is conveniently possible. For an accumulation of $N$ photons there will be an ordinary statistical uncertainty of $\sqrt{N}$ photons. Moreover, there is the contamination of the background sky to contend with. It adds photons all over the place. The main problem now becomes one of discerning the star unambiguously against this statistically uncertain background of photons.

The large telescope and long observation time still aid in detecting the star, but the gain is discouragingly small because the collection of stellar photons is accompanied by a proportional collection of confusing sky photons. The real gain to be had is by isolating the stellar photons from the sky photons. At high magnification the sky gets spread out and thereby diluted with respect to the star, which remains approximately pointlike. It is that isolation that is a major function of the telescope. Ideally, larger telescopes not only should be able to gather more photons because of their larger collecting areas, but they also should provide superior isolation because of having higher potential resolving power as limited by diffraction. Atmospheric turbulence intervenes, however, smearing the image so as to limit resolution to about a second of arc, regardless of the potential diffraction limit, which may be up to 50 times finer. That topic will be dealt with in a subsequent chapter. In the meantime, the emphasis is that, even for spectroscopy, light buckets should be unacceptable as telescopes.

If the minimum resolvable angle is $\phi$, then the sky background flux in one resolvable picture element (pixel) is proportional to $B(D\phi)^2$, where $B$ is the sky brightness within the wavelength band and $D$ is the diameter of the telescope. The uncertainty, or noise, is proportional to the square root of that amount. The stellar flux is proportional to $D^2$, and including observing time $t$ as a parameter leads to a signal-to-noise ratio of

$$S/N \propto (D/\phi)\sqrt{(t/B)}\ .$$

Note that the minimum observable angle $\phi$ is just as important as the telescope diameter $D$ in achieving the ability to detect faint stars. As was already mentioned, that "seeing" angle is typically about a second of arc. Considerable effort is expended to find sites where $\phi$ might be the smallest, and sometimes even greater effort is expended to site observatories above the offending atmosphere.

Although the above relationship has a certain applicability, we should remain wary of applying it indiscriminately. There are sufficiently frequent situations, especially in spectroscopy and narrow-band imagery, where the uncertainty is dominated by the signal statistics, in which case $\phi$ does not enter and we are concerned solely with collecting as many photons as possible. Whichever the situation, ultimately there just is no substitute for photons. That means as large a diameter as feasible, and even more so if the observing time is limited by schedule or by a transient nature of a phenomenon.

## Modest Telescopes

On the other hand, recent years have lead to a crying need for more observational facilities to outfit the throngs of young astronomers. So far, telescopes have been localized at a few prime but relatively inaccessible sites and are controlled by a few major consortia. This means that young astronomers have to go through much bureaucratic red tape, often to no avail, just for the opportunity to observe for themselves. Furthermore, even senior astronomers who happen to be out of favor may be denied observing time, and prolonged contiguous observing is extraordinarily difficult to come by. The remedy, it seems to me, would be a proliferation of modest-sized telescopes in the 1.5- to 2-meter class that are sufficiently standardized for economy. The size is large enough to accomplish significant work, yet it is small enough to become affordable if the design were standardized. All too frequently small institutions insist upon designing their own telescopes almost from scratch to fulfill their individual tastes. It should be evident that that course is hopelessly uneconomic, just as it would be uneconomic for each automobile to be custom designed. In essence, what is needed for observational astronomy is numerous standardized telescopes at somewhat suboptimal sites that are much more accessible to these young astronomers. Perhaps it is just my own taste, but nevertheless there does appear to be a need for relatively inexpensive telescopes in the 1.5- to 2-meter class. Even at this modest size there does not remain much question between alt-azimuth (alt-az) and equatorial mounting; computers now make the former more economical. Alt-az has the further advantage of simplifying the primary mirror support since tilting is restricted to the elevation axis. So far, however, no one has taken up the cudgel in favor of asymmetric alt-az mountings. Figure 1 shows the interior of what I have in mind.

The offset of the azimuth axis confers several advantages. First, the support of the telescope yoke is relegated to a small thrust bearing at the pintle on one side and inexpensive air-pads on the other side. Economically, the track for those air pads also can be used for independently supporting the mailbox style building as shown in Figure 2. The offset also provides a large comfortable working area at the focus. The shutter can be a single piece that would slide to a position over that

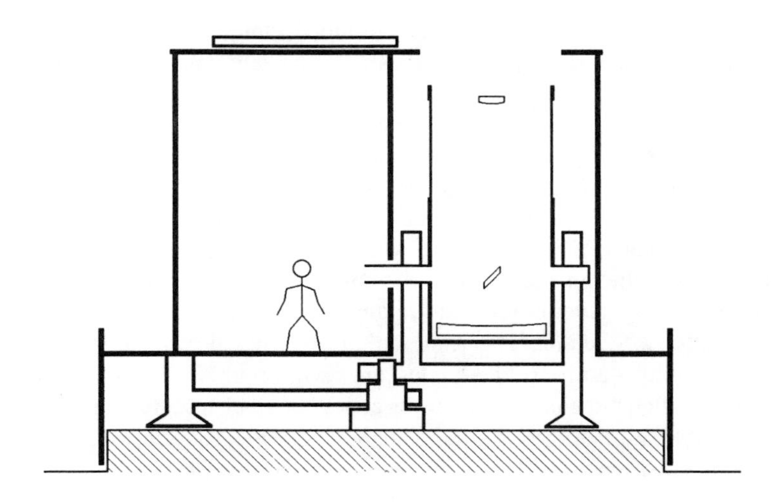

FIGURE 1. Asymmetric alt-az telescope.

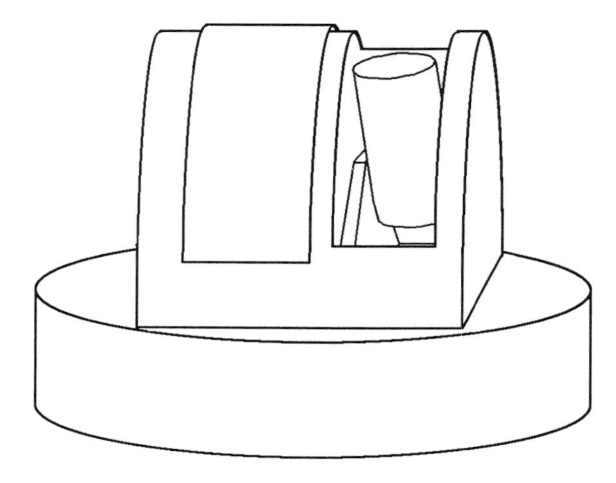

FIGURE 2. Dome.

area so as not to be vulnerable to the wind when it is open for observing. The single piece is certainly less expensive and easier to seal against inclement weather than a multipiece shutter.

Whatever design might be elected, however, it should be versatile enough to cope with the general variety of observational chores.

## Large Telescopes

Even though the need for larger numbers of modest telescopes is broader, there is always the urge for surpassing the biggest. The present competition started with the Mount Wilson 60", then their 100", to be followed by the Mount Palomar 200", now called the Hale 5-meter telescope. Those telescopes each dominated their respective eras. The quest is now being faced with the problem of monolithic versus segmented apertures. There are projects in the United States and Germany to make 8-meter mirrors. These mirrors are so large that active control of their support is essential to maintaining their figure within tolerance. I have three reservations about the monolithic approach. First, the approach is more and more difficult to scale up for future generations. That is already an impasse, so that when these 8-meter telescopes are complete, they will not hold the crown for size, although they will argue that arrays of these telescopes will reign. Well, why not arrays of somewhat smaller, more manageable telescopes? Second, 8-meter mirrors are very fragile and difficult to transport. If you are not extremely careful, you may end up with a segmented aperture whether you like it or not. Third, these mirrors require a large aluminizing chamber. For practicality, the chamber would have to be located near the telescope on the mountaintop site. There the chamber sits as a large capital investment that only gets utilized maybe once a year and so is economically quite inefficient.

The segmented approach actually has a fairly long history of attempts that never achieved regular operational status, such as the 4.2-meter telescope of Pierre Connes at Meudon pictured in Figure 3. The primary of that telescope was composed of 36 mirrors, each having a spherical figure and 60-centimeter square aperture, for the collecting area of a 4-meter telescope. The cassegrain secondary was figured to correct the spherical aberration, and the motive of the telescope was high-resolution Fourier spectroscopy of rather bright sources. Because of the brightness of the sources, the optical alignment could be referred to the light of the source itself. Even though it never became regularly used, that telescope served as a forerunner learning exercise in that it showed that the coalignment of mirror segments could be controlled adequately and that air pads could provide suitable support for the azimuth.

The first segmented aperture telescope to achieve significant use was the Multiple Mirror Telescope (MMT) atop Mount Hopkins in Arizona. Its configuration is six 1.8-meter telescopes on a common mounting and organized to bring their light to a common focus. As such, it might more properly have been called a multiple- telescope telescope. Apart from its observational achievements it demonstrated several design successes. It showed the utility of alt-az mounting and that a

FIGURE 3. Segmented-mirror telescope at Meudon (courtesy P. Connes).

boxy rather than hemispheric dome did not aggravate the seeing. Furthermore, the structural design, based on finite-element analysis, rescued the telescope from an awkward failure. There had been a laser system to maintain the accurate coalignment of the six telescopes. This system failed as a result of moths flying into the laser beams. The fallback procedure has been to coalign the telescopes on a nearby bright star, and then to lock the alignment and offset to the desired direction. Thanks to the rigid structural design, the coalignment remained valid for about half an hour. Although that procedure would have been considered unacceptable in the initial specifications, it served quite successfully. On the other hand, it must still be a nuisance since the plan now is to replace the six telescopes with a single telescope having a greater collecting area. My major objection to the MMT concept is its lack of scalability. The problem is that the six cannot be increased to a much larger number.

The latest size crown is held by the 10-meter Keck telescope. The segments of this telescope are figured as off-axis pieces of a parabola and have a hexagonal outline. Fabrication of the segments uses a technique called stress-mirror figuring, whereby the mirror blank is bent by weights and then is figured to a spherical surface. When the weights are released the figure springs back to that of the off-axis parabola. An initial problem was that the stressing could only be done for a spherical-outline blank, and when the blank became trimmed to a hexagonal outline, residual stresses slightly upset the figure. This problem may be fixed temporarily by warping harnesses that appropriately bend the mirrors while in the telescope. More recently, ion milling has been used to touch up the figure before mounting in the telescope. My main objection to all of this is that the resulting telescope is not inexpensive.

The coalignment of the segments is based on capacitive edge sensors. To the best of my knowledge, their success is the first time that aperture segments have been maintained actively to astronomical tolerances without reference to the starlight itself.

## Fixed Spherical Primary

Nevertheless, all of these endeavors have reaffirmed my longstanding conviction that the most economical way to construct a really large telescope is to use a fixed spherical reflector, just like the Arecibo radio telescope. In 1980, Aden Meinel presented a graph showing the cost per unit area as a function of diameter for various optical and radio telescopes. I have added several telescopes that he omitted to Figure 4. Note in particular that the Arecibo radio telescope is meritoriously so far removed from the herd of radio telescopes that it lies outside the original boundaries of the graph. At the same time, my own estimated cost for a fixed spherical primary optical telescope places it vectorialy displaced from the herd of optical telescopes, the same as the Arecibo case and for basically the same reasons.

The spherical figure brings a multitude of virtues. For example, when segmented, all of the segments have the same figure, and, furthermore, that figure is far and

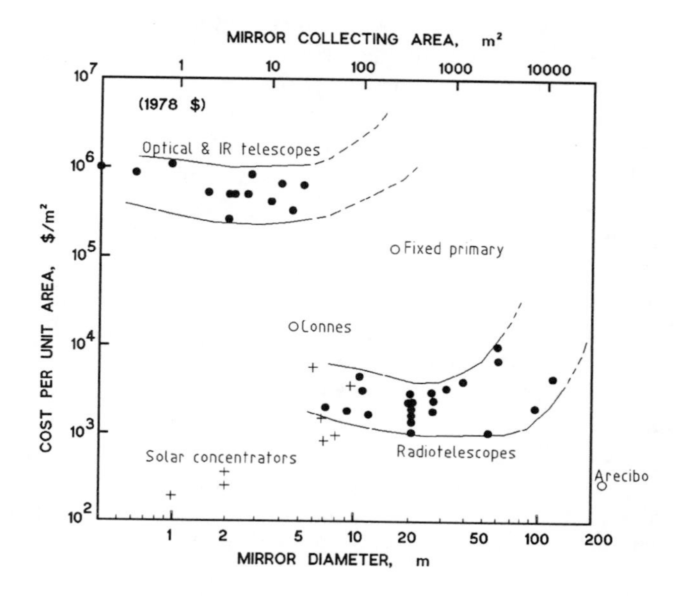


FIGURE 4. Relative costs of telescopes (adopted from A. Meinel).

away the easiest to make, test, and coalign. For those reasons, the mirror surface is far less expensive per unit area and the segmentation can be to thousands, rather than the small numbers for other schemes. Large numbers gain even further economies from mass production.

Cost estimation of individual mirrors can be done by expressing their cost as a polynomial function of diameter. There will be a constant fixed handling cost per mirror followed largely by the third order, the cost per unit weight. Dividing both sides by the square of the diameter gives the cost per unit area. This function has a rather deep minimum that occurs when cost attributable to weight is twice the cost per unit. If we empirically fit a curve to published prices of astronomical mirrors, we find that the minimum is indeed rather deep and occurs at about 0.2 to 0.25 meter diameter, a conspicuously small size. For that reason it is most economic to have many fairly small segments. On the other hand, that must be balanced with the necessity for more mounting and alignment fixtures, and so my preference would be about 0.75 meter size for the segments.

Another very important, if not the most important, feature of the spherical figure is that it has no axis; therefore the primary reflector may remain fixed, while steering is accomplished by moving only the much smaller and lighter auxiliary mirrors. A fixed primary would even do away with any need for active alignment, since almost all of the alignment problems arise from variations of gravitational loading that result from steering. However, there is a penalty in that more primary mirror surface is required because not all of the area is used at one time. This penalty depends somewhat on the sky coverage desired, but for reasonable coverage the factor is about six. The enormous savings that accrue from the spherical figure would far more than make up for that penalty factor.

Since the concept for a fixed spherical primary never acquired the favor of any sponsorship, the best way for me ever to see how the overall engineering might fit together has been to construct a few models. Figure 5 pictures the most elaborate and corresponds to a 13-meter effective aperture, although the actual intention is for 15-meter effective aperture.

The mirror support structure is basically a geodesic frame built from very many identical aluminum castings, and the mirrors themselves are from diamond-machined aluminum blanks, probably with minor touch-up polishing. The main criteria for the alloy selection should be long-term stability after annealing and freedom from inclusions. It is quite inexpensive to diamond-machine the mirror surfaces since numerical control is unnecessary for generating spherical figures. It is also easy to polish spherical figures if the diamond-machined finish should need touching up. The casting shapes are sketched in Figure 6.

The geodesic form is a regular icosahedron inscribed in a sphere, where the faces of the icosahedron are filled with a triangular lattice and puffed out so that the vertices of that lattice also lie on the sphere. We are concerned only with about a 120° portion centered on one of the icosahedral faces. The three vertices of that face will be specially made to serve as the three- point support for the entire structure, and the rim also will have to be specially made quite rigidly to avoid warping. Although the triangular latticework necessarily involves an assortment of slightly different edges and angles, the identical castings shown in Figure 6 allow for the slight differences by having elongated holes for their lap jointing and by being thin enough to flex slightly near the hub of the casting. The mirror castings have a stress relief design so that their spherical figure will not get deformed from the simple three-screw adjustable attachment. The outlines of the mirror castings will be hexagonal, but trimmed to be slightly irregular so as to minimize the gaps between neighboring mirrors.

The optical configuration is shown in Figure 7. The geometric optical details will be discussed a little bit later. For the moment, suffice it to say that, except for the primary reflector, all of the components are mounted on a common frame. That alt-az steerable frame is supported in azimuth on three air pads. The frame must be held in accurate position with respect to the center of curvature of the fixed primary reflector. Here again the spherical figure of the primary is a blessing because if we have a point light source adjacent to a four-quadrant photodetector, and that light source is imaged at the center of the detector, then we know that the center of curvature of the reflector lies exactly halfway between the source and the detector. Simply to avoid obstructing the telescope beam with paraphernalia at the center of curvature, that center is folded downward with a small flat mirror. Now, if the folding flat is bent just a little to induce a small amount of astigmatism at 45° to the quadrants of the detector, then the difference signal of the diagonals of the quadrants also serves to measure focus in addition to the normal $x$-$y$ sensing

FIGURE 5. Model of Arecibo-style optical telescope.

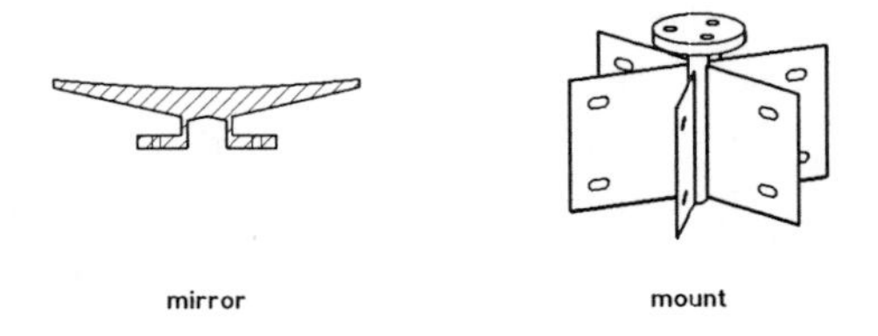


FIGURE 6. Casting shapes.

of the quadrants. That copies the clever servosystem ordinarily used for compact discs.

The shape of the dome lies somewhat in between the traditional hemispheric and the box of the MMT. Although the resemblance is more to a battleship gun-turret, the architectural form is fairly æsthetic. That is important since telescopes are meant to be inspiring. For this 13-meter version the dome diameter is about that of the Hale 5-meter telescope while being only about half as high. Much more importantly, the alt-az arrangement of the telescope allows an internal bracing structure within the dome so that it can be much lighter. (A fully hemispheric dome can be both light and strong, but when a slot is required the dome must become very much heavier to be adequately strong.) The internal structure includes walls so that there is a very comfortable, large observing room where the telescope focus conveniently arrives through one of the walls. There is even room for a visitor's gallery above the observing room, and an elevator provides access to those rooms from the ground.

The dome has a relatively short single-piece shutter which, as was mentioned earlier, is economical and easy to seal against inclement weather. The aperture spanned by the shutter is large, so that an exoskeleton is recommended for strength.

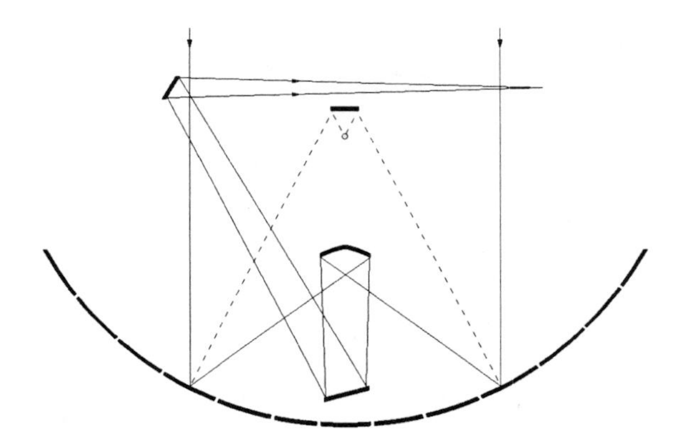

FIGURE 7. Optical configuration.

The dome itself is supported on air pads that share the same track as those that support the secondary mirror structure. That track is effectively the roof of an annular base building. There are two levels of rotating seal between the dome and the base building, and those two levels are separated by a corridor containing trolley-wire electrical power connections to the rotating dome. That corridor is restricted for maintenance, since open trolley wires present a hazard.

Any such large and radical telescope certainly should be preceded by a small-scale prototype. In this case, a 2-meter version would be about the smallest that might be really helpful. In fact, I did build a corresponding model prior to the model of the 13-meter version, and that prototype model strongly influenced many details of the adopted design.

## Optical Design

The major objection to a spherical primary mirror comes from optical aberrations, mainly spherical and coma. These are especially serious with the very fast focal ratio ($f \approx 0.6$) being considered here. The original incentive for designing this telescope was as a light collector for Fourier transform spectrometry; and in that case only one sharp pixel is required, so only the spherical aberration required correction. However, it shortly became clear that such a large telescope must be more versatile with a capability to produce good images that have many pixels. These demands led to my development of two geometric design procedures for the correction of arbitrary spherical aberration and coma. Zero spherical aberration occurs when the optical path length (OPL) for all rays is exactly the same. That is Fermat's Principle. Zero coma occurs when the rays satisfy the Abbé sine condition, whereby the sines of the ray angles through the focus are proportional to the initial ray heights at the entrance aperture. Thus all rays have the same image scale. Failure in that regard is called *offence* against the sine condition (OSC). When both of these aberrations have been corrected the optical system is said to be *aplanatic,* and the significance of such correction will be discussed subsequently. Treatises on optical design contain many equations, and good optical designers feel quite comfortable among those equations. On the other hand, there are those of us who feel quite uncomfortable amongst them, and so the procedures that I will describe are based more on geometric graphical concepts. It seems that this attitude productively complements the more traditional approach.

The first procedure is based on the simple optical behavior of ellipses that will be used to correct only the spherical aberration of a spherical mirror, or for that matter the spherical aberration of any ensemble of rays. By ray I mean not only a line in space, but also a direction along the line and a fiducial point on the line that specifies an instant of propagation time. A second point could be specified farther along on the ray. The points may be regarded as the tail and head of an arrow. Somewhere in between, however, the ray may be bent by reflection so that the second point will no longer lie on the line of the original ray. The idea of getting the light ray from the first point to the second suggests reflection at an elliptical surface. The first and second points are the foci of an ellipse, and the time delay or path length of the ray arrow is equal to the major axis of that ellipse; knowing the positions of the foci and the major axis completely specifies the ellipse. The next question is where on the perimeter of the ellipse does the reflection take place? If the equation for the ellipse is expressed in polar coordinates about the initial focus, then the locus of the reflection is obtained directly from the direction of the initial ray. The polar equation for an ellipse should be well known as that for a Keplerian orbit to anyone having taken astronomy. It is

$$r = \frac{a(1 - e^2)}{1 + e \cos \phi} ,$$

where $a$ is the semimajor axis and $e$ is the eccentricity obtained from $2ae = f$, where $f$ is the distance between foci. Figure 8 shows two ways to apply this concept to solve for a secondary reflector to correct the spherical aberration of a spherical primary mirror.

The idea is to bring an incident ray starting at H to an eventual focus at F via reflections at P and S such that the overall pathlength L is specified. The ellipse, shown dashed, osculates the secondary reflector at its reflection point S. The left-hand version chooses P and F as the foci of the osculating ellipse, and the distance HP must be subtracted from the overall path to get the major axis of that ellipse. The right-hand version chooses B and F as the foci, and the distances HP and PB must be subtracted from the overall path L to get the major axis. Both versions arrive at exactly the same secondary reflector, and the spherical aberration of the primary is completely nullified at F. The program for either version amounts to about a half-page of FORTRAN code, and that for version A is in Appendix 1. Required input information is radius of the primary, whose center of curvature serves as origin of the coordinate system, maximum entrance ray height, final focus position, and optical pathlength from the $y$-axis to that focus.

The procedure allows designing the secondary mirror for the configuration shown in Figure 7. Previous solutions to that problem by others have been more

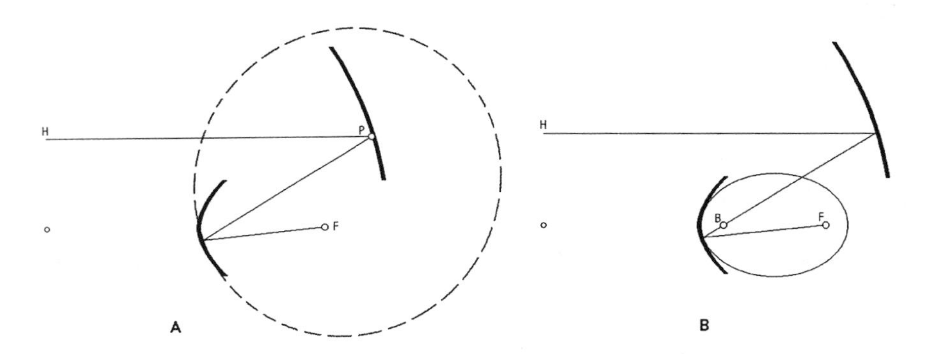


FIGURE 8. Geometries of osculating ellipses for calculating a secondary mirror that corrects spherical aberration. Geometry A forms the basis of program TELES in Appendix 1.

mathematical and less graphic. The final two folding mirrors of the configuration are flat, and so they do not contribute aberrations.

A curious design results when the final Gregorian focus (without any folding flats) is assigned to be coincident with or near the paraxial prime focus of the spherical primary. The secondary, as shown in Figure 9, looks like the inside of a wine glass. Notice that the rays converge on the focus from a solid angle of more than $3\pi$ steradians. If the configuration were used as a solar collector, the equilibrium temperature at the focus then would be hotter than that at the solar surface. That may seem odd, but remember that the solar surface sees a hemisphere of cool sky whereas the focus sees flux from almost all around. The configuration also would have made a superb searchlight, but the need for searchlights has practically vanished after World War II. The design also might be convenient for laser-driven inertial-confinement fusion since the focus gets impinged from so many directions.

The design of Figure 9 also can lead to an excellent microwave or millimeter wave antenna. Such a configuration already has been arrived at previously by von Hoerner (and perhaps a few other designers) using analytical techniques, and he recognized a special polarization problem. Were it not for polarization, the pattern at the focus of Figure 9 would provide an excellent match to that of a dipole oriented along the optic axis. The problem is that, when the reflectors unfurl the pattern, the polarization becomes radial over the aperture. Opposite sides of the aperture then cancel. Even though it is possible to design a wire antenna that would properly match the focal pattern, at the high radio frequencies involved wire antennae are too lossy. A better solution is to place a conical splash plate over the end of a waveguide feed as shown in Figure 10. Since the size of the splash plate is on the order of a wavelength, the details of its shape should take diffraction into account.

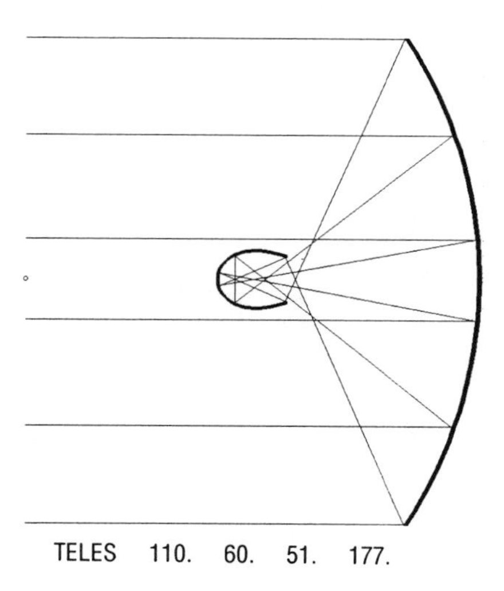


FIGURE 9. Collector with spherical primary corrected for spherical aberration. Light comes to the focus over more than $3\pi$ steradians.

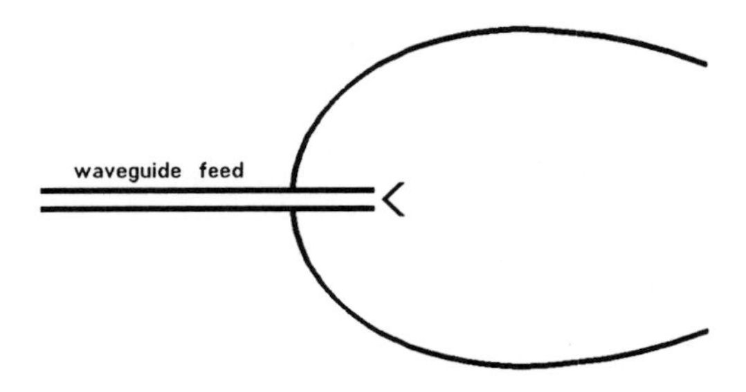


FIGURE 10. Matching a waveguide to Figure 9 with a splash plate.

The real motive for such a configuration is quietness. The feed is practically in a Faraday cage and so sees almost no spurious illumination to contaminate the signal. That quietness should be very important for sensitive radio telescopes. For that reason, I would have thought this design preferable to the one actually selected for the upgrade of the Arecibo radio telescope, the upgrade being intended to receive more bandwidth than the original distributed line feeds. For radio telescopes in general, the configuration also offers other practical benefits such as ease of fabrication, alignment, and metrology. Sometimes, as with Arecibo, a fixed primary is tolerable; but when horizon-to-horizon coverage is required for aperture synthesis, a steerable primary is in order. Even with a steerable primary, though, this design still could be better than an orthodox parabolic reflector.

Moving the final focus even closer to the primary's center of curvature leads to the configuration of Figure 11. It is most unorthodox in that it presents a drastic

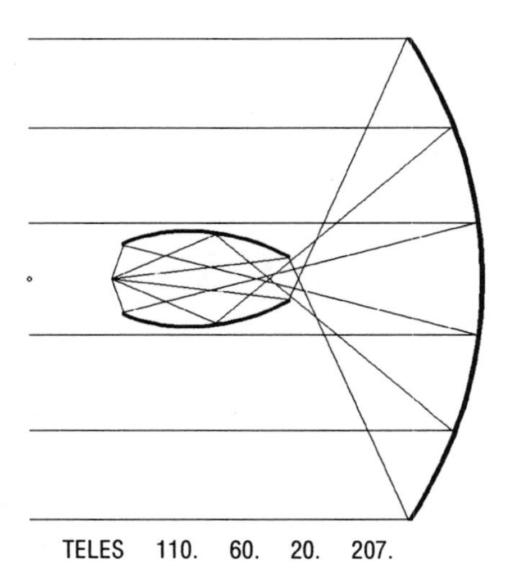


FIGURE 11. Strange corrector for spherical aberration.

violation of Abbé's sine condition. We will concern ourselves much more about that condition a little later on. In this case, the sines of the emerging rays are not only not proportional to the sines of the incident rays (the entrance ray heights), but the outermost incident rays have the smallest angles onto the focus. The question is whether that is of any importance for a point source matching to a waveguide feed. I do have a vague recollection of such a configuration being published (probably in the 1950s) as a proposal for a radio telescope, but I am unable to track it down and nothing seems to have come of it. It is not at all clear whether the proposal was discarded for the sine problem, because it was too unorthodox, or for simple lack of interest.

While the telescope configuration of Figure 7 is sufficient for the original spectrometric application, it fails to produce images at all. The reason for this failure is that the various zones of the aperture, even though they all focus to the same point, they do so with very different magnifications. Thus only the single-axial pixel stays sharp. Figure 12 shows effective focal length as a function of zone, and we can see that the inner zones end up with almost five times the magnification of the outer zones, so the discrepancy is very severe. The extreme severity of this coma results from the very fast focal ratio under consideration.

It turns out that there are a variety of ancillary optical systems that can correct the coma. A cleverly simple scheme that is based on a concept called *hemisymmetry* was suggested to me by Paul Robb. Imagine a 1 : $N$ scale replica of the telescope placed back to back with the telescope, such that their foci coincide, and insert a field lens such that the aspheric mirrors of each are conjugate. This arrangement is like the hemisymmetry described by Conrady, except that we have a focus rather than a pupil at the center of hemisymmetry. The arrangement constitutes an $N$ power afocal telescope where the height of the exit ray is rigorously proportional to the height of the entrance ray to give uniformly the same magnification $N$ for all zones. The afocal telescope then could be used in conjunction with a small good-quality camera to provide reasonable images.

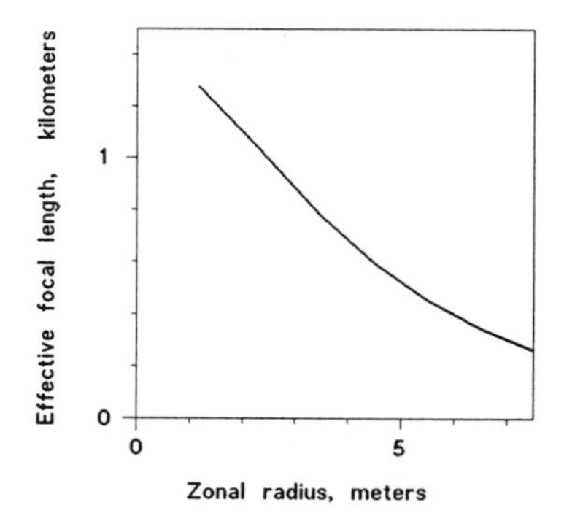


FIGURE 12. Zonal variation of image scale for Figure 7. This is really severe coma.

Now it would be nice to dispense with the camera by incorporating its power into the telescope replica. Also, shifting the field lens toward the main telescope avoids the smallest asphere being uncomfortably small while keeping the overall length reasonable. The shifted field lens then also acts as a focal reducer to magnify the convergence angle of the focus and shift the focus toward the telescope. These modifications make the correcting optics appear as in Figure 13 and sufficiently perturb the hemisymmetry so that the larger mirror of the corrector must be slightly deformed to accomplish the coma correction. When both the coma and the spherical aberration are nullified, the system is said to be *aplanatic*.

The necessary deformation of the larger mirror in Figure 13 is estimated as follows. Reverse the propagation direction of Figure 13 so that it looks more like Figure 7 and adapt the calculation so that the light starts from the point as compared with the plane wavefront of Figure 7. The deformations from a spherical figure of the corrector's larger mirror, positions, and pathlengths all are judiciously adjusted until its magnification as a function of zone (analog of Figure 12) becomes a close match with Figure 12. Although this attempt can give substantial improvement, the match to Figure 12 persistently remains imperfect with disappointing performance.

## Dihedral Remedy

The exasperation of trial-and-error groping for a match led me to a more basic approach that abandons any reliance on hemisymmetry and goes directly to the problem. The task is to take any given set of rays, such as those emerging from the focal point of the main telescope, and transform them into another set of rays having preassigned directions through a subsequent focal point while preserving optical pathlength (OPL) invariance.

The key to the reparation is based on the simple properties of a dihedral mirror. The double reflection from a dihedral mirror, as shown in Figure 14, ends up deviating a ray by twice the angle of the dihedral. Thus you can wobble the dihedral module without wobbling the direction of the output ray. Furthermore, if the dihedral pivots about its vertex, the output ray does not shift at all, not even with

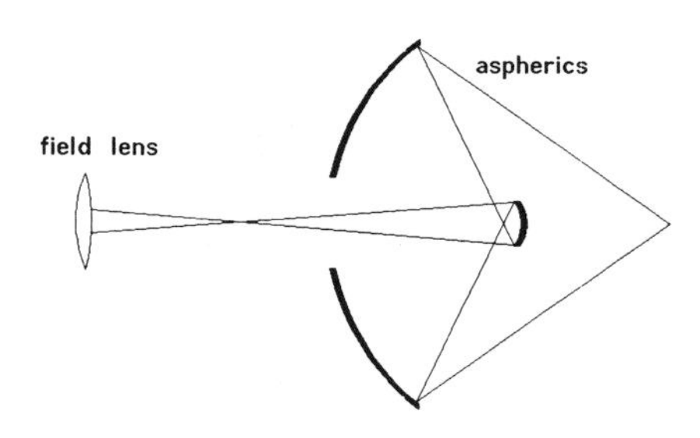


FIGURE 13. Coma correcting module for Figure 7.

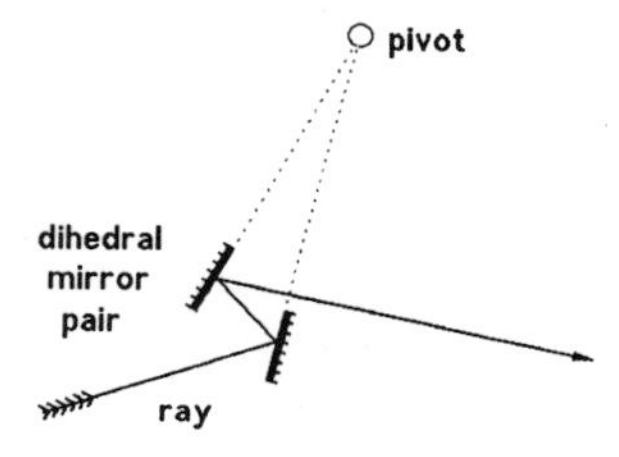


FIGURE 14. Reflection by a dihedral. Recognize that the result is invariant with respect to rotation of the module about the pivot.

respect to optical pathlength. Now suppose that we take an input ray segment of length PL and wish to rotate it to an output ray segment of the same pathlength as shown in Figure 15, where the segments are depicted as archery arrows having heads and tails. Just where would we place a hinge point such that the input arrow can be rotated to the position of the output arrow?

Figure 15 shows the geometric solution. The hinge or pivot point P is at the intersection of the perpendicular bisector between the tail positions and the perpendicular bisector between the head positions. Rotating either arrow about P can move it to the position of the other arrow. That rotation can be accomplished by establishing a dihedral mirror with vertex P and having the appropriate dihedral angle. For example, if one mirror intercepts a head or tail, the other mirror should lie on the adjacent perpendicular bisector. The dihedral then can take any rotational position about P without affecting the ray transformation. Each rotational position offers a pair of reflecting points, one on the input ray associated with one on the output ray.

Having accomplished the problem for one pair of input-output rays, the next step is to do it for a neighboring pair of input- output rays such that the reflecting facets describe smooth continuous surfaces. One way to do this is shown in Figure 16, where two neighboring rays are drawn. From a reflecting place R1 on one

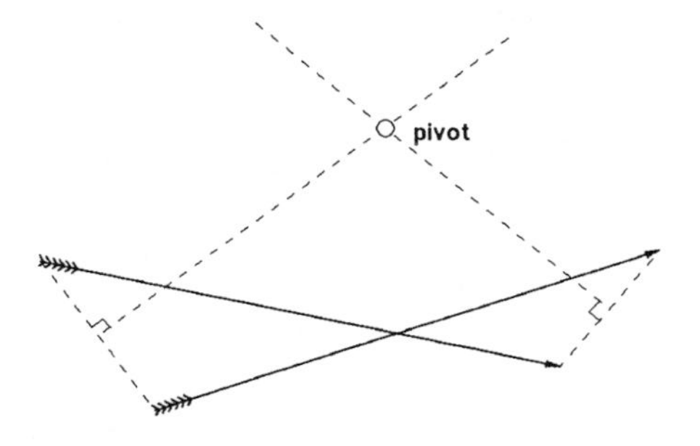


FIGURE 15. Finding the pivot for the dihedral reflective transfer of ray arrows.

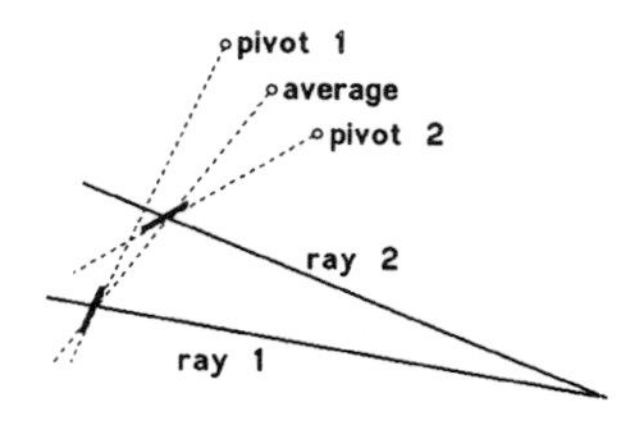


FIGURE 16. Extrapolating the mirror surface from one ray intercept to the next.

ray, where its mirror surface must go through its pivot point P1, we want to get smoothly to a reflecting place R2 whose mirror surface will go through P2. A reasonable way to get there is to draw a line through R1 and the average of P1 with P2. The intersection of that line with the second ray gives R2.

Another way might be to draw two lines, one from R1 to P1 and the other from R1 to P2. The average of their intersections with the second ray then would be R2. This second route is slightly lengthier computationally, but it may be easier if the procedure is expanded to three dimensions, because in that case the pivots are themselves lines and the average of two lines is awkward.

The programming for this is all very straightforward and simple, being based on the elementary analytic geometry of straight lines, perpendicular bisectors, and intersections. Program MPLAN for solving two-mirror aplanatic transfer systems is given in Appendix 1. It solves mirror shapes for transferring an ensemble of input rays to an ensemble of output rays. The program could be easily modified to accept a comatic input ensemble of rays for designing correctors such as those in Figure 13.

## Microscope Objectives

Without modification, the program designs aplanatic all-reflecting microscope objectives. Typically, these have a concave primary and a smaller convex secondary like a Cassegrain telescope, and for the special circumstances of a Schwarzschild microscope objective their figures are essentially spherical. For different input parameters, however, the program generates extremely unusual objectives such as the one shown in Figure 17 whose numerical aperture almost equals the refractive

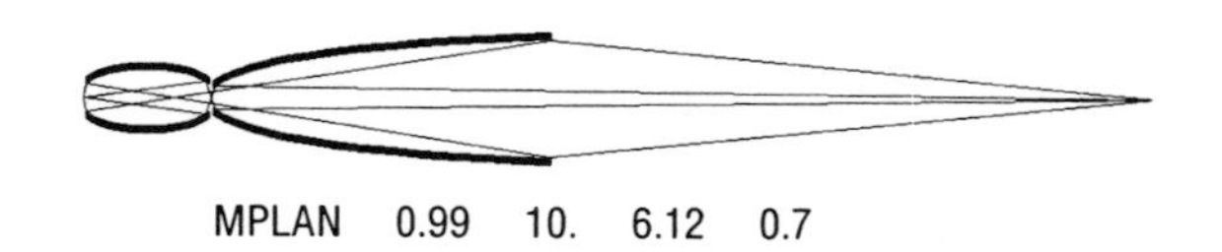

MPLAN 0.99 10. 6.12 0.7

FIGURE 17. Fast aplanatic microscope objective.

index of the operating medium. If its magnification were reduced to unity, the configuration would become two congruent prolate ellipsoids having colinear major axes and sharing a common focus.

The program also designs more conventional configurations such as Schwarzschild microscope objectives. These are rather like Cassegrain telescopes, except that the source is up close and the obscuration of the secondary is a real nuisance. That nuisance might be avoided by offsetting the secondary to give an off-axis system such as the one shown in Figure 18. While an off-axis version of the dihedral procedure program could design such systems, all we have to do is recognize that a spherically concave primary creates a rather good virtual image of the source, and the convex elliptical secondary transfers that virtual image to a magnified real image. The off-axis feature imparts some astigmatism, which could cured by a slight bending of either mirror.

Aside from possible application in soft X-ray lithography, another interesting prospect for Figure 18 would be dark-field illumination. You could simply shine a laser or other highly collimated light straight down through the hole in the primary for reflective illumination, or in the opposite direction, up toward the hole, for transmissive illumination. The interesting aspect is that now the illumination has low numerical aperture while the imaging has high numerical aperture commensurate with high resolution. This would be a beneficial reverse of conventional dark-field systems such as cardiod condensers.

## Telescope Correction

Getting back to the task of aplanatic correction for spherical primary reflectors, the program MPLAN in Appendix 1 is easily modified so that the tails of the ray arrows emerge not from a point source but from the surface of the spherical

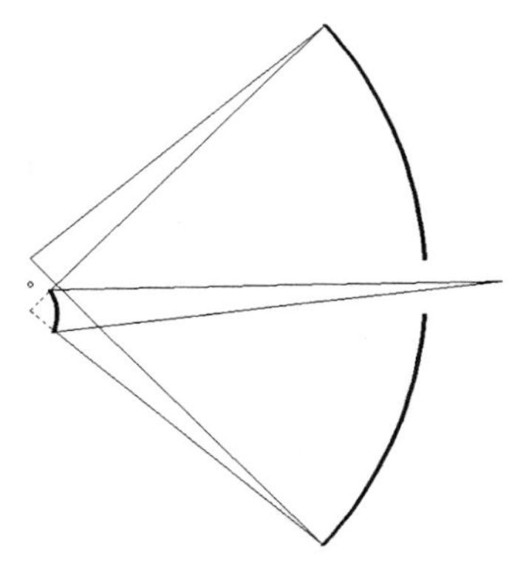

FIGURE 18. Off-axis reflecting microscope objective.

primary mirror. The lengths of the ray arrows are also diminished by the distance from the $y$ axis to that mirror, just as was done back in Figure 8A. Then, instead of using the osculating ellipses, we proceed with the dihedral technique. The resulting modified program RPLAN also is given in Appendix 1.

An unexpected and curious result from the progam has been strange telescope designs that are nevertheless aplanatic. Figure 19 shows my favorite example that might find application as a large infrared telescope because of easy baffling so that the detector would see nothing but the source or easily cooled surfaces. Axially symmetric odd-mirror telescopes such as this must revert their images, but because of the axial symmetry the reversion must take place in depth. Thus the focus moves in rather than out for foreground objects.

Figure 20 shows several other unusual telescope designs that came out of the program in Appendix 1. I think that the same procedure was used to design the corrector optics of the Spectroscopic Survey Telescope (now called the Hobby-Eberley Telescope), although I am not fully aware of the details.

The unusual nature and variety of these designs present a challenge for optimization. Clearly there are numerous solutions to the coma problem for telescopes having audaciously fast spherical primary reflectors. The problem now is to evaluate the designs with optimization in mind, and this probably can be best done with off-axis ray tracing. For that end I have been adamant in urging the development of a skew ray-tracing progam for parametrically described surfaces. There exist many skew ray-tracing programs for surfaces described by explicit aspheric equations and also for splines, but such descriptions are not suitable for reentrant surfaces, some of which appear among the designs. For these surfaces it is more appropriate

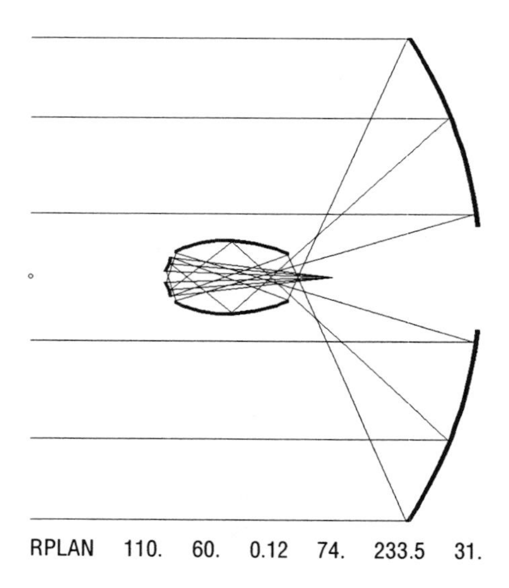


FIGURE 19. Interesting aplanatic telescope.

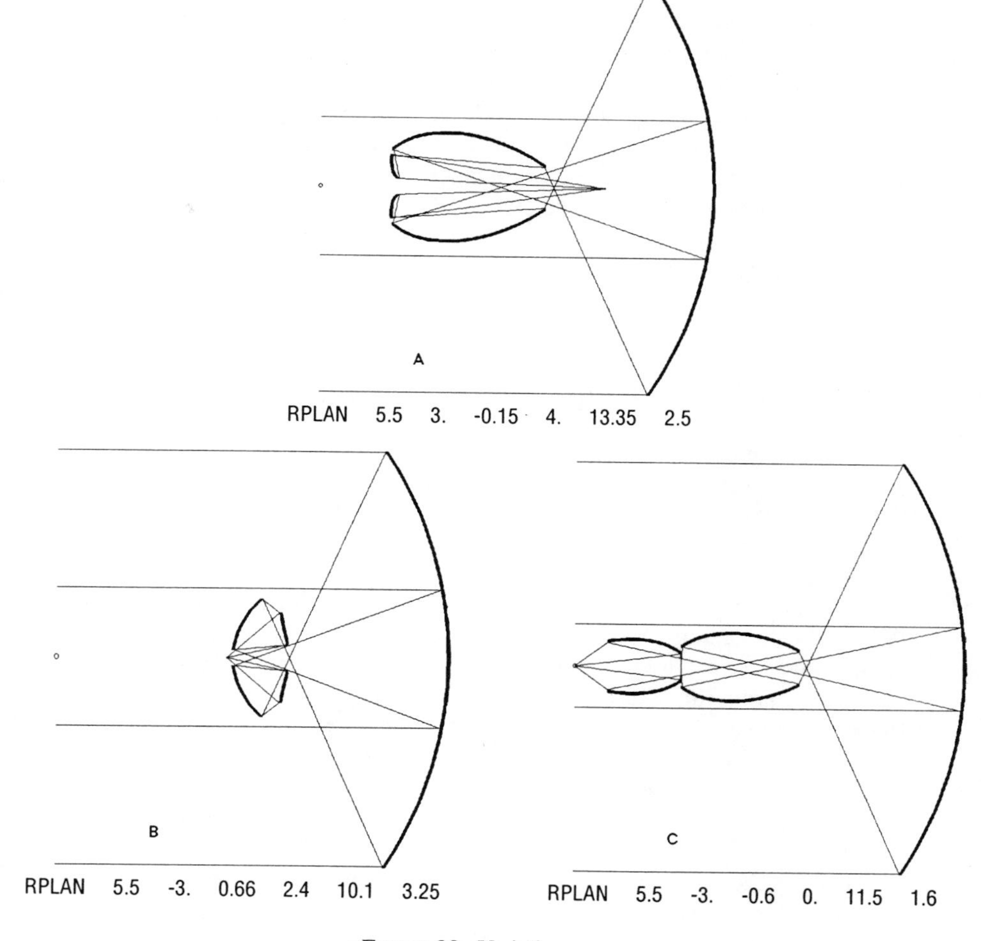


FIGURE 20. Variations.

to express the mirror surface coordinates, $x$ and $y$, locally as separate functions of a third parameter, such as the corresponding initial entrance height $H$. This parametric form is expressly that which is provided at the the output of the given design programs. Computer graphics experts do have programs for ray-tracing parametric surfaces, but it is not fully clear as to whether these are applicable or to the precision required. I myself have neither the experience, the perseverance, nor the temperament to write such a program, and so the design evaluations and optimizations will remain unknown until someone else tackles the problem.

Once it does become possible to evaluate the off-axis performance of these systems, it should be possible to optimize them. Furthermore, one might relax the Abbé sine condition or change it to a tangent or angle proportionality or the Herschel sine condition that will be described shortly. Similarly, the spherical aberration might be slightly relaxed. These alterations would be difficult for analytic as opposed to geometric design schemes. It is known that often for optical systems, small changes of design can lead to significant changes of performance. The other side of the coin is that tolerances can become very stringent, but the practices of metrology and control have improved so much over the past few decades that stringency may not be too onerous. As examples, all of the Keck telescope segments are controlled to very tight tolerances, very large scale integrated circuits have become a reality, and compact discs are commonplace.

## Alternative Prescriptions

Remember that coma gets cured by satisfying the Abbé sine condition. Not only does that satisfaction provide for uniform magnification, but it also transports the illumination uniformly to the focus. Radio telescope designers often refer to that uniformity or to the tapering of illumination, but it is really the same concept as satisfaction of the sine condition.

When the rays converge over more than $2\pi$ steradians, such as back in Figure 9, the Abbé sine condition loses its meaning. It is then more appropriate to consider the Herschel sine condition, whereby the sine of one-half of the ray angle through the focus is proportional to the incident ray height on the entrance aperture. The Herschel sine condition tolerates focal ray angles up to 180° and maps entrance solid angles uniformly to exit solid angles.

From a qualitative standpoint let us briefly consider the opportunity for correcting the Figure 7 telescope directly, rather than with an accessory system like that of Figure 10. The three small mirrors all are available for figuring, but it is probably best for the secondary to retain rotational symmetry for ease of fabrication. However, that secondary need not be designed specifically to correct spherical aberration. Although it is quite speculative, it might be possible to calculate the secondary so as to distribute the illumination uniformly on a plane normal to the axis and near where the tertiary intersects the axis. The behavior then may be somewhat as if that plane were a conjugate image of the entrance pupil. The main function of the tertiary and the final mirrors then would be to readjust the ray angles toward the final focus, rather than also having to redistribute the illumination. Ideally, the dihedral procedure could be expanded to three-dimensional form to handle the tilted mirrors, but a reasonable approximation can be had simply by considering the plan and the perpendicular views separately. It is expected that advanced technologies, such as the stress-mirror figuring that was used for the Keck segments, might be able to fabricate those off-axis tilting mirrors.

The off-axis versions of the dihedral also can lead to the design of ultrashort unobscured aplanatic telescopes with spherical primaries as shown in Figure 21. The secondary mirrors of such designs resemble tablespoons and so may be too

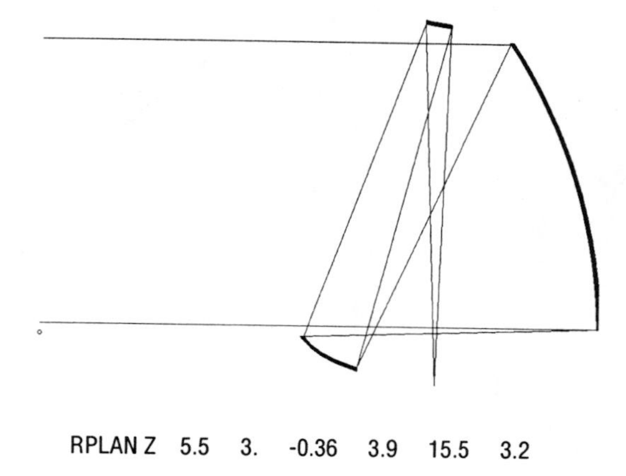


FIGURE 21. Ultrashort unobscured aplanatic telescope with spherical primary.

difficult to fabricate. Moreover, it is likely that their fields would be too small for most applications.

## Geometric Variants

A unit-magnification version of Figure 17 would be simply a double ellipsoidal aplanatic transfer system. Such a system also could be folded and rotated as shown in Figure 22A, and then the input focus could be shifted to infinity, resulting in the strange three-mirror telescope having vanishingly small central obscuration such as that seen in Figure 22B. Although it is no longer strictly aplanatic, the

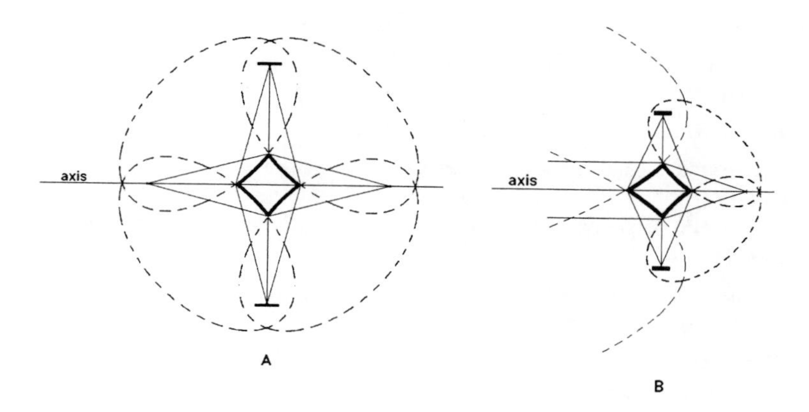


FIGURE 22. Folding and rotating ellipses and parabolas.

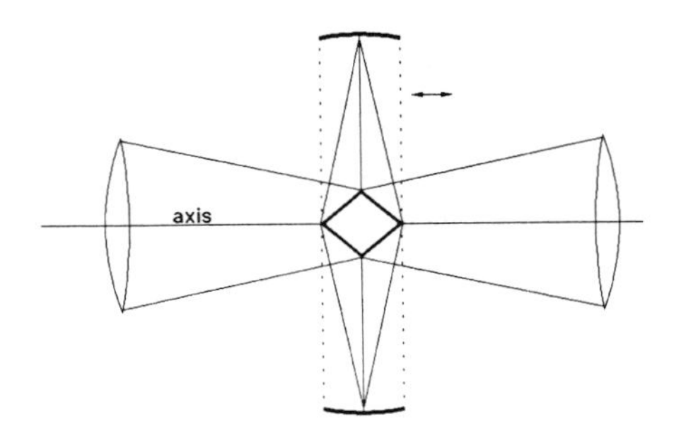


FIGURE 23. Adjustable coma corrector.

surface figures could be easily altered slightly using the dihedral procedure to restore aplanaticity.

In conclusion, Figure 23 shows a variant of Figure 22A using purely conical axicons and a spherically figured belt mirror that all could be fabricated and joined as a plexiglass disk. It would serve as an image transfer system where the coma could be adjustably corrected by translational shifting of the disk between the lenses.

## Bibliography


J. Chevillard, P. Connes, M. Cuisinier, 1977 "Near infrared astronomical light collector" *Appl. Opt.* 16: 1817

V. Galindo, 1964 "Design of dual-reflector antennas with arbitrary phase and amplitude distributions" *IEEE Trans. Ant. Prop.* AP-12: 403–408

A. K. Head, 1958 "The two-mirror aplanat" *Proc. Phys. Soc. London* 70: 945–949
———"A class of aplanatic optical systems" Proc. Phys. Soc. London 71: 546–551

A. Hewitt, *ed.*, 1980 *Optical and Infrared Telescopes for the 1990s,* Kitt Peak National Observatory

D. Korsch, 1989 *Reflective Optics,* Academic Press

L. Mertz, 1981 "Geometrical design for aspheric reflecting systems" *Appl. Opt.* 18: 4182–4186
———"Aspheric potpourri" *Appl. Opt.* 20: 1127–1131 E. Popko, 1968 *Geodesics,* U. of Detroit Press

C. Wang and D. L. Shealy, 1993 "Differential equation design of finite-conjugate reflective systems" *Appl. Opt.* 32: 1179–1188

G. D. Wasserman and E. Wolf, 1949 "On the theory of aplanatic aspheric systems" *Proc. Phys. Soc. B* 62: 2

# 2

# X-ray Telescopes

## Introduction

The spectra of stars, even for early-type (hot) stars, fall off steeply in the ultraviolet. The falloff is roughly according to the Wien radiation law that prevails on the short-wavelength side of the black-body radiation peak. Nevertheless, there has been considerable interest in the ultraviolet because many spectral features pertaining to ionized atoms lie in that region, and so help in studying the stars. Because our atmosphere becomes quite opaque to hinder the study of ultraviolet and X-rays, they held a high priority for incipient space astronomy. Then, during a rather modest rocket mission to look for X-rays from solar wind or cosmic bombardment of the moon, the discovery of a powerful point-like souce in the constellation Scorpius came much to the surprise of the astronomical community. That discovery heralded much more to come, all of which begs for X-ray telescopes. The problem is that materials do not refract or reflect X-rays, so that we cannot just extend optical designs to the X-ray region. The lack of reflection is a slight overstatement since grazing incidence reflection can be had, and often is used for soft X-rays.

An alternative to direct imaging is indirect imaging by means of shadowcasting. Shadowcasting is particularly suited to the X-ray region because opaque materials are available and the wavelengths are so short that diffraction is negligible, leading to sharply-defined crisp shadows. There are several strategies to be described, each of which has its own attributes.

## Coded Aperture Imaging

I derived the concept that is now called *coded aperture imaging* from the looks of Gabor holograms back in 1962. It is vexing for me to have to stake the claim myself, but most X-ray astronomers, if they acknowledge any origin at all, mistakenly credit Dicke [1968] and Ables [1968], who later came up with the concept independently.

A Fresnel zone pattern originally served as the coded aperture. By replacing a camera lens with a coarse Fresnel pattern reticle as shown in Figure 1, the ensemble of shadows from stars mimics a Gabor hologram. While the shadowing itself may

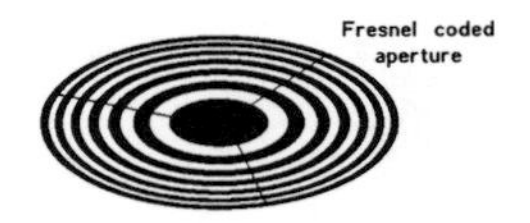


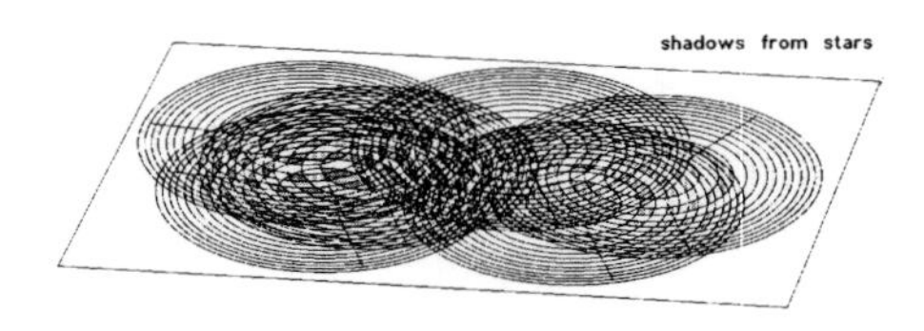


FIGURE 1. Coded-aperture imaging using a Fresnel pattern.

be from X-rays, an optical reconstruction from a minified copy of the hologram then reproduces the star images. The image formation has been broken into a two-step process. My colleague, Niels Young, concocted certain techniques to improve the image reconstruction by eliminating unwanted diffraction orders, and in our 1962 publication we indicated that random variable patterns also could also be used in principle even though the simple optical reconstruction then would not apply. Young's idea to use off-axis zone patterns led to low-noise optical reconstructions, but now computers excel for that job.

The key ingredient is that the coding pattern have an autocorrelation consisting of a sharp spike on an innocuous background. The Fresnel coding pattern functions as a two-dimensional analog of the chirp in chirp radar, which had just become public at about the same time. A variety of other patterns, such as Hadamard sequences and uniformly redundant arrays (URAs), fulfilling the autocorrelation properties have since come in to fashion; but other than style, they do not offer a fundamental advantage over the Fresnel pattern.

A Fresnel pattern, as shown in Figure 2, is composed of alternating opaque and transparent rings. The radii of the boundaries are at the square roots of successive integers, so that each ring occupies an equal area. Small versions serve as diffraction lenses, and larger versions generate fascinating moiré fringes. The reader is encouraged to make a transparency copy of Figure 2 (as well as Figures 4 and 5 to come) for seeing these moiré fringes. By exactly superposing two matched patterns, the pair transmits half the light, the same as for a single pattern. If one member of the pair is slightly displaced by the width of an outermost ring, then one full cycle of moiré fringe appears across the patterns. That full cycle of decorrelation confers an independence such that when the pattern pair functions as a diffractive lens two point foci become resolvable. In the context of coded-aperture

FIGURE 2. Fresnel zone pattern.

imaging, the angular resolution (in radians) is then the width of the outermost ring divided by the distance from the coded aperture to the film. Similarly, for patterns other than Fresnel, the resolution is the width of the finest features of the pattern divided by the distance from the aperture to the film.

The next consideration is sensitivity. Here the reference comparison will be a pinhole camera whose pinhole size is the diameter of the smallest feature of the coded aperture, and whose spacing to the film is the same as for the coded aperture. This pinhole camera would have the same resolution as the coded aperture. Clearly, the coded aperture gathers far more photons than the pinhole in proportion to their respective open areas. For an isolated star there is a signal-to-noise (S/N) improvement according to the square root of that area ratio. Overlapping of the shadow patterns contaminates that improvement, so that for a continuous picture there is no improvement on the average. The result is that pixels that are brighter than the rms level win an S/N improvement, while those that are fainter suffer a degradation. The detection of faint X-ray stars then would be severely hampered. However, that degradation is not at all necessary. It is merely a consequence of poor linear reconstruction that overlooks the useful knowledge that the image must be entirely nonnegative. Improved nonlinear techniques that overcome that degradation will be considered later in this chapter, but failure to appreciate their potentiality unfortunately has already discouraged much use of shadowcasting schemes.

Since the telescope must operate in space, because the atmosphere is opaque to X-rays, an awkward deficiency of the scheme is film recovery. Furthermore, film does not make a very sensitive detector for X-rays. Photon counters are the preferred detectors, and they must have positional sensitivity, such as imaging proportional counters or CCD detectors. That preference over film seriously constrains the possible trade-offs for the telescope design parameters of aperture size, angular resolution, and angular field. We can have a fairly large area imaging proportional counter with rather large pixels or a small area CCD with commensurately small pixels. So, if we want high angular resolution over a small field, we end up with a ridiculously long instrument or a pathetically small aperture. That is why all of the coded-aperture telescopes that have been flown have had rather coarse angular resolution.

## Moiré Telescopes

Here, the systematic elegance of the Fresnel zone pattern, as contrasted with the pseudorandom patterns, can be called in to play. Imagine now a second pattern identical to the Fresnel coded aperture as an exit aperture immediately in front of the detector. An X-ray star will cast a shadow of the entrance aperture through the exit aperture, and the result is parallel equispaced moiré fringes. The spacing and orientation of those fringes are sensitively dependent on the star location. The large detector pixels now need only resolve the coarse moiré fringes rather than the fine Fresnel rings to achieve high angular resolution. The transformation from a point (star) to such fringes amounts to Fourier transformation, and so the image should be recoverable via inverse Fourier transformation.

It is not quite as simple as that, however. The fringes are not ideal Fourier components in that they have a dc bias and their phases are not quite right. The dc bias results from intensity detection that cannot be negative. This bias problem can be overcome by taking a difference signal from complementary systems. The phase problem is twofold. Inasmuch as the image is not necessarily symmetric, the sine and cosine components are independent and must be detected separately. Furthermore, the phase origin for the fringes does not stick on the axis of the telescope, but sits halfway between the centers of the exit zone pattern and the shadow pattern. The situation calls for multiple telescopes, as in Figure 3, and is mathematically represented by expanding the kernal of Fourier transformation (for clarity, only the one-dimensional formulation is given here):

$$\begin{aligned} F(x) &= \int f(y)\exp(i2\pi xy)\,dy \\ &= \exp(i\pi x^2)\int [f(y)\exp(i\pi y^2)]\exp(-i\pi[x-y]^2)\,dy\,. \end{aligned}$$

The expansion shows that Fourier transformation is equivalent to premultiplication

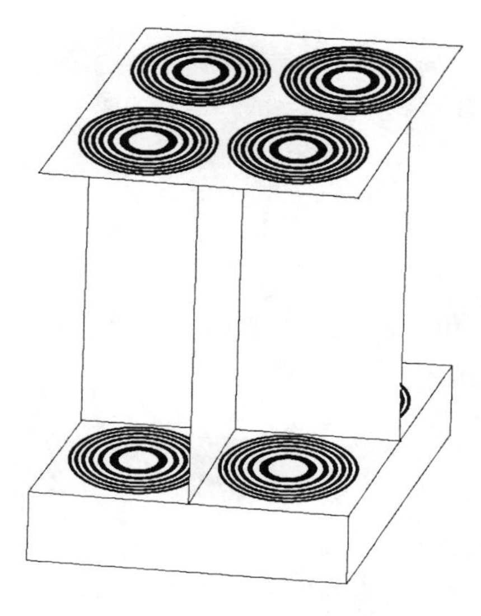

FIGURE 3. Configuration for a moiré telescope.

by a chirp, $\exp(i\pi y^2)$, followed by convolution with a chirp, followed by postmultiplication with a chirp, $\exp(i\pi x^2)$. This one-dimensional expansion has come to be known as the *chirp-z formulation* of the Fourier transform.

In terms of the moiré telescope, the premultiplication has been omitted, the shadowcasting of the entrance pattern gives the convolution, and transmission through the exit pattern gives the postmultiplication. The result (in one dimension) amounts to Fourier transforming $[f(y)\exp(-i\pi y^2)]$ rather than $f(y)$, so that $f(y)$ later can be recovered by multiplying the output transform by $\exp(i\pi y^2)$. That final multiplication amounts to shifting the phase reference for the moiré fringes to the center of the exit zone pattern.

Nevertheless, because of the complex formulation it is necessary to have separate detections for the cosine (real) and sine (imaginary) parts and to eliminate the dc bias on each so that one ends up with a four-composite system as shown in Figure 3 for a properly general moiré telescope. Furthermore, we have yet to mention how to get sine as opposed to cosine moiré fringes. The solution for this was proposed and published by Rogers [1977]. He introduced an odd form of the Fresnel pattern where the radii of the zone boundaries are proportional to the square roots of odd integers rather than even integers. Figure 4 shows his odd form, which should be compared with Figure 2. Each form also has its complement having a dark center. Now the moiré of an even-odd or an odd-even pair gives sine fringes, whereas the moiré of an even-even or odd-odd pair leads to cosine fringes. Adjustable-phase moiré fringes can be had with a spiral pattern as shown in Figure 5, where rotating one member of the pair with respect to the other changes the phase of the fringes.

FIGURE 4. Rogers' odd form of Fresnel zone pattern. Compare with Figure 2.

By the way, the advent of computers and high-resolution raster graphics printers, based on lasers or inkjets, makes it quite easy to draw all of the forms of Fresnel patterns. In the $xy$ raster, simply make the pixel black or white according to a selected bit of the integer binary expression $(x^2 + y^2 + s\phi)$, where $\phi$ is the desired phase and $s$ is a scale factor sized to change the selected bit with a $\pi$ phase change. The spiral pattern is had by letting $\phi = \arctan(y/x)$. Numerous variations on the pattern theme turn out to be quite curious and entertaining. For example, you can change the $+y^2$ to $-y^2$ to get a hyperbolic zone pattern, or you can change the sign to an exclusive-or or other logical relation to get a boxy zone pattern.

## Rotation Modulation Collimator

In 1965, Minoru Oda described a modulation collimator. It consisted of two identical equispaced parallel grids situated on the ends of a tube. Transmission through the tube falls immediately on a simple detector (i.e., having no position sensitivity). An X-ray star would cast a shadow of the entrance grid onto the exit grid. Motion of the star with respect to the tube would move the shadow across the exit grid to modulate the X-ray transmission to the detector; hence the name *modulation collimator*. Oda had the tube pointing out normal to the spin axis of a rocket, and

FIGURE 5. Spiral Fresnel zone pattern.

the intention was mainly to see if the X-ray source in Scorpius was smaller or larger than the projected grid spacing.

It soon occurred to me that if Oda's collimator were pointed along the spin axis, as shown in Figure 6, rather than normal to it, then any star in the field would undergo a distinctive fm modulation pattern dependent on both coordinates of the star's position. The number of modulation cycles per turn would be proportional to how far off-axis the star was, and the phase of the fm would reveal the star's azimuth, as is shown in Figure 7. These modulation patterns are so distinctive that even a superposition of many of them might be solved by correlation to reconstruct a picture of the stars. The considerable advantages with respect to coded-aperture or moiré telescopes are that there is no need for a position-sensitive detector and that a spinning, rather than a three-axis stabilized, vehicle can be used.

The mathematical formulation for the patterns of Figure 7 is

$$T(\theta) = \frac{1}{\pi} B \operatorname{abs}[\pi - \operatorname{mod}_{2\pi}(X \cos\theta + Y \sin\theta + \delta)] ,$$

where $B$, $X$, $Y$ are the brightness and coordinates of the star, with the brightness adjusted to account for vignetting, and $\delta$ is a phase alignment constant. An

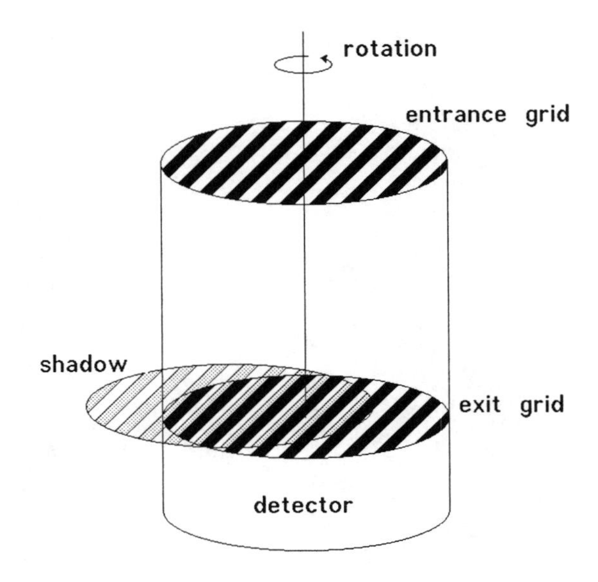


FIGURE 6. Rotation modulation collimator.

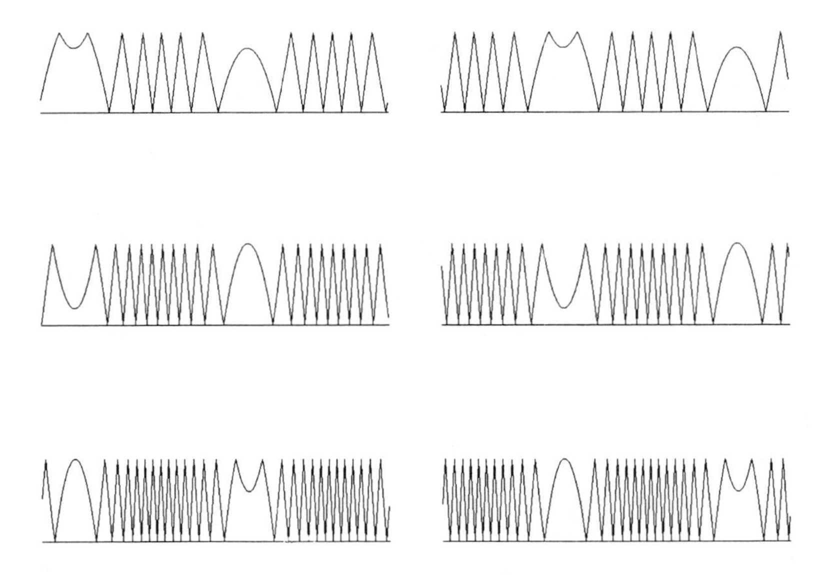

FIGURE 7. Modulation patterns.

acceptable sinusoidal version is

$$S(\theta) = \frac{1}{2} B [1 + \cos(x \cos\theta + Y \sin\theta + \delta)] .$$

In analogy with chirp radar, an appropriate name for these periodic fm modulations is *warbles*.

For a few stars in the field the scheme actually worked in practice. The first rocketborne experiment by Schnopper *et al.* [1970] gave the positions of several X-ray stars near the galactic center to within about 4 arcminutes, or a factor of about 20 better than prior devices. That is appreciable because it reduces the area to be scoured for identifications by a factor of 400.

Nevertheless, a serious problem with the side lobes becomes evident. This problem is more easily interpreted by adopting a Fourier point of view. The modulation collimator is sensitive to an equispaced pattern of parallel fringes projected on the sky. Strictly speaking, the fringes are triangular, as shown in Figure 7, rather than sinusoidal, but mechanical imperfections tend to blur the sharp peaks and for practical purposes the fringe pattern is quite like a Fourier component in the same sense as the moiré fringes described earlier. Rotation of the collimator then provides some information on a circle, and only on that circle, in the Fourier transform domain of the picture. As a result, the inverse transform gets a point spread function of $J_0(x)$, rather than the $[J_1(x)/x]^2$ Airy disk of ordinary telescopes. The side lobes of the Bessel function $J_0(x)$ are horrendous; alternate lobes not only go negative, but each lobe contains as much energy as the central lobe.

Before proceeding any further we should take note of certain qualities of the warbles shown in Figure 7. These warbles appear qualitatively similar twice per rotation. The similarity is not necessarily exact and depends on the precise orientation of the spin axis with respect to the projected bar patterns. What we do not want is for one half of the rotation to give information that is redundant with that from the other half, because in the reconstruction that easily gives rise to artifact images separated by 180° in azimuth from the true star images. The same problem was encountered for nonsymmetric images in the moiré telescope, wherein we needed independent sine and cosine information. In principle, that information can be available from differences and sums of the 180° opposite information, if extraordinary care is taken in the orientation of the spin axis with respect to the projected bar pattern of the collimator. Specifically, the bar patterns must represent sine plus cosine functions ($\delta = \frac{1}{4}\pi$) for that to work successfully. Those acquainted with the Hartley transform will recognize sine plus cosine as cas functions. However, it is likely to be very risky to assure the orientation of the spin axis with adequate precision.

A strongly recommended alternative is to fly the collimators in pairs, with one slightly shifted to give sine fringes with respect to cosine fringes of the other. The shift ensures that their signal information is nonredundant regardless of spin-axis offsets. In the mathematical formulations above, the shifts are identified with $\delta$. The collimator pair becomes analogous to a radio-astronomy antenna pair where they are so accustomed to providing a quadrature signal channel that it is taken for granted. Now the versatile realm of complex mathematical analysis becomes applicable with the quadrature signals serving as the real and imaginary parts of a single complex variable.

In the Fourier spirit whereby a sum of $N$ sinusoids plus a constant is sufficient to depict exactly an arbitrary set of $2N + 1$ equispaced real samples, we may ask whether it is possible to find $N/4$ (or some comparable specific number of) warbles whose sum exactly depicts $N$ complex samples. This is asking for a sparse picture containing that number of stars that would give exactly the observed data, and the trick is to find the locations and brightnesses (perhaps complex) for those stars. Over the years I have made several thrusts at solving this problem. Although they all have been futile, for some the reasons remain inexplicable or not fundamental, so a brief description of their outlines follows.

For the first approach, the composite signal is supposed to be dominated by that of the brightest star. Assuming data from a collimator pair, as was recommended, the sinusoidal version of the complex signal composed from $k$ stars and with the dc removed is

$$S(\theta) = \sum_k B_k \exp i(X_k \cos\theta + Y_k \sin\theta + \delta) .$$

This sum of warbles is a vector sum, and if one warble is strong enough, it should dominate the phase of the composite vector, allowing us to solve

$$\phi = \arctan \frac{\mathrm{Im}\{S\}}{\mathrm{Re}\{S\}} .$$

The phase $\phi$ is incomplete in that it is only the principal value and must be unwrapped to be of benefit. Ordinary unwrapping can be done with the recursive filter,

$$\Phi_\theta = \Phi_{\theta-1} + (\phi_\theta - \Phi_{\theta-1})_{\pm\pi} ,$$

where the $\pm\pi$ subscript signifies that the parenthetical expression is reduced to the interval $\pm\pi$, just as is done for the evaluation of trigonometric functions. Successive values of the unwrapped phase $\Phi$ necessarily differ by less than $\pi$, and the fractional part of $\Phi_{\theta-1}$ cancels out, leaving the proper fractional part $\phi_\theta$. If all goes well, then we can calculate a least squares solution of

$$(X \cos\theta + Y \sin\theta + \delta) = \Phi_\theta$$

for the $X$, $Y$ coordinates of the dominant star in the picture. Once the coordinates are known, the brightness can be found from the original signal equation by correlation. Thereafter, the signal contribution from that dominant star can be subtracted, and we can look for the next dominant star. The big problem comes from unreliability of the phase unwrapping. When there are more than a few stars in the field, even in the absence of noise, the fainter stars easily conspire such that the phase unwrapping fails, slipping cycles. Erroneous star positions result, and the whole scheme breaks down. In the next chapter it will be found that the inclusion of a time constant in the phase unwrapping equation makes it much more tenacious and reliable. The improved equation would introduce phase shifts and amplitude modifications, albeit predictable, that would alter the apparent star positions and frankly has not been tried in the present context.

An attempt for a closed-form solution noticed that the Fourier series expansion of a warble is

$$F(n) = B\,(\exp 2\pi i\phi)\; J_n(R)\;,$$

where $R$, $\phi$ are the polar coordinates of the star position. At one time I had presumed to introduce a complex position $Z = R\cos\phi + iR\sin\phi$ to give for a composite signal

$$F(n) = \sum_k B_k\, J_n(Z_k)\;.$$

Although intricate, and possibly ill-conditioned, that would have led to a closed-form solution for all $Z_k$ and $B_k$ in terms of Neumann series. There is no point in belaboring the mathematical details because in fact the presumption is incorrect and the problem remains unsolved.

Recognizing that the side lobe problem arises from the sparcity of our information in the Fourier domain, we might hope via Cauchy's theorem to deduce information interior to the circular locus from information on that locus. If $f(z)$ is a function of $z$, analytic at all points on and inside a contour $c$, then $\oint_c f(z)\,dz = 0$. Interpreting the $f$ as the Fourier transform of the image and the contour as our circle of information, we then might be able to deduce the value of $f$ at points interior to the circle using Cauchy's integral formula,

$$f(a) = \frac{1}{2\pi i}\oint_c \frac{f(z)}{z-a}\,dz\;.$$

"This remarkable result expresses the value of a function $f(z)$ (which is analytic on and inside $c$) at any point within a contour $c$, in terms of an integral which depends only on the value of $f(z)$ at points on the contour itself" [Whittaker & Watson 1915, p. 89]. A preliminary numerical simulation to test this approach attempted to infer the interior of the Fourier $U$, $V$ domain information for simple fringes corresponding to a single point in the original image. The test did not appear successful and was not pursued further, even though programming errors were a distinct possibility. It remains a shame to dismiss the strategy without understanding the reason for failure.

Yet another approach sought to ensure the nonnegativity of the image. Before worrying about nonnegativity, we first must ensure that the image should be purely real. If we denote by $\pm$ subscripts the quadrature collimators, then we can form a guaranteed Hermitian signal that has no dc bias by applying the recipe

$$H(\theta) = \{T_+(\theta) - T_-(\theta) + T_+(\theta-\pi) - T_-(\theta-\pi)\} \\ + i\,\{T_+(\theta) + T_-(\theta) - T_+(\theta-\pi) - T_-(\theta-\pi)\}\;.$$

An additional feature of this recipe is that the sharp peaks of the triangular modulation waveform are truncated, resulting in a trapezoidal waveform that more closely resembles the sinusoidal version.

The nonnegativity of the source implies that the circle of information in the Fourier domain is not only Hermitian, but also that it is part of an autocorrelation function. Let us proceed to construct an antecedent function $G(\theta)$ on a circle congruent to that of $H(\theta)$. Although $G$ will be zero except on that circle, its autocorrelation will be distributed over an area extending out to twice the radius of the circle and will agree with $H$ on the circle supporting $H$. The autocorrelation of $G$ may be visualized graphically as shown in Figure 8, by displacing a complex conjugate copy of $G$ and summing the products of the two intersections to give

$$H(\theta) = G\left((\theta + \frac{\pi}{3}\right) G^*\left(\theta + \frac{2\pi}{3}\right) + G\left(\theta - \frac{\pi}{3}\right) G^*\left(\theta - \frac{2\pi}{3}\right) .$$

The normalization of $G$ due to its being nonzero only on an infinitely narrow ring need not be considered here because discrete samples of $H$ and $G$ are used. The equation is underdetermined for finding $G$, and so we can arbitrarily choose $G$ to be Hermitian as well as $H$; $G(\theta) = G^*(\theta + \pi)$. That choice leads to a set of simultaneous equations for $H(\theta + \frac{\pi}{3})$, $H(\theta)$, and $H(\theta - \frac{\pi}{3})$ whose solution is

$$G(\theta) = \sqrt{\frac{H(\theta + \frac{\pi}{3})H(\theta - \frac{\pi}{3})}{2\,H(\theta + \pi)}} .$$

The square root again presents underdetermination, so we can successively choose the samples of $G$ so as to minimize the magnitude difference between each sample and its predecessor:

$$\text{minimize} \mid G(\theta_n) - G(\theta_{n-1}) \mid ,$$

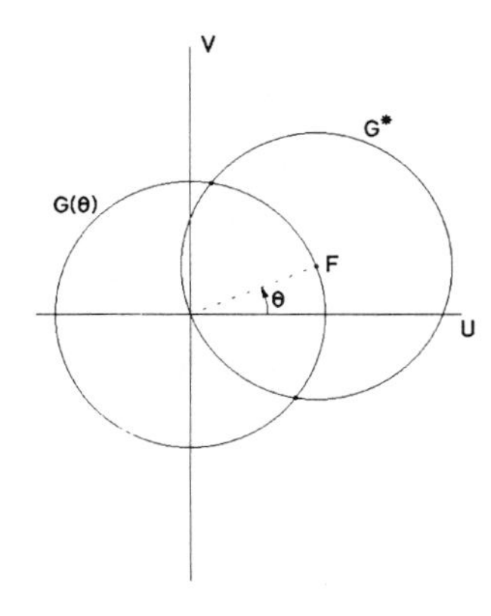


FIGURE 8. Graphic construction of the autocorrelation $F$ of a function $G$ lying on a circle.

initializing $G(\theta_0)$ to have a positive real part. For a single star the resulting point spread function becomes $J_0^2(r)$, whose side lobes decay more rapidly than those of $J_0(r)$. The procedure does not seem to give proper results when more stars are present. Because of all the underdetermination, there are numerous variations on this same theme of constructing an antecedent function. Although it seems that they should work in principle, they apparently do not in practice.

## Rotational Aperture Synthesis

Inasmuch as none of the approaches for side lobe reduction appeared satisfactory, the recourse chosen is to actually measure the information not just on a single circle in the Fourier $U$, $V$ domain, but on many concentric circles in that domain. This is accomplished by using many collimator pairs with a large variety of grid pitches. Figure 9 shows how the entrance grid pattern might appear for an array of nine collimator pairs. Sampling the measurements at 64 azimuth angles then provides information at all of the intersections of the sampling lattice in the $U$, $V$ plane, as shown in Figure 10. Although that sampling density is excessive near the origin, it is just about right to give approximate squares near the periphery. For each of the nine radii there will be an $H(\theta)$ which may be labeled as $H(R, \theta)$. These values are interpreted as samples in the Fourier $U$, $V$ plane even though they are begotten from triangular rather than sinusoidal modulation. The next step is to reconstruct an image from those samples. Although faulty, Fourier inversion offers the most straightforward procedure for reconstruction. Simulations of the whole

FIGURE 9. Aperture patterns for image synthesis from nine pairs of rotation modulation collimators.

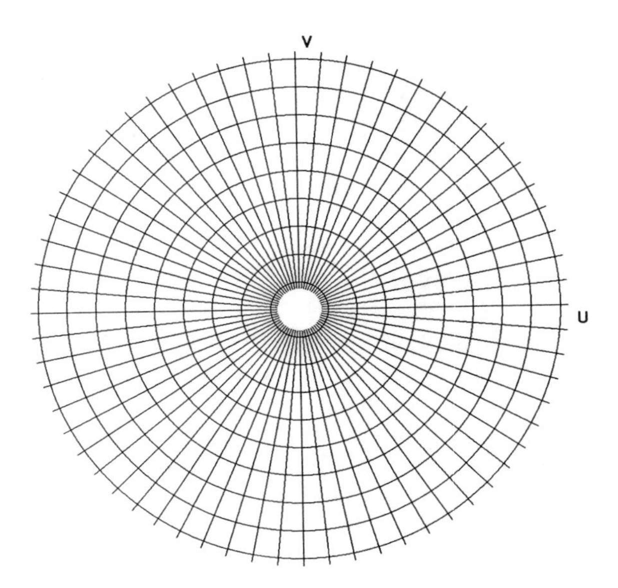


FIGURE 10. Sampling lattice in the $U$, $V$ plane for the apertures of Figure 9.

process will not only show its efficacy and limitations, but also its susceptibility to quantum statistical noise. The FORTRAN source code for the simulations is given in Appendix 2.

The first step in the simulation is to form the signal to be expected from $k$ stars in the field,

$$T_{\pm}(R, \theta) = \frac{1}{\pi} \sum_{k} B_k \operatorname{abs}\left[\pi - \operatorname{mod}_{2\pi}\left(RX_k \cos\theta + RY_k \sin\theta - \frac{\pi}{2} \pm \frac{\pi}{4}\right)\right],$$

where $B_k$, $X_k$, $Y_k$ are the brightness and coordinates of the $k$th star, with the brightnesses already adjusted to account for vignetting within the nominal field of view that is circumscribed by $X^2 + Y^2 = \pi^2$. The $\pm$ refers to the collimator pairing, $R$ is the number of bars across the field of view, and $\theta$ is the azimuthal orientation of the array. The seemingly curious phase shifts $-\frac{\pi}{2} \pm \frac{\pi}{4}$ serve for prescribing a simple addition-subtraction Hermitianizing recipe that converts $T_{\pm}(R, \theta)$ to $H(R, \theta \pm \pi)$, which assures that our Fourier transform reconstruction will be purely real.

Classical Fourier synthesis is more suitable to the polar sampling lattice of Figure 10 than Fast-Fourier-Transform algorithms. Further considerations in favor of classical synthesis are that the output image is purely real and that we would like an excessively sampled rendition of that image. Many of the multiplications nevertheless can be avoided by making secondary lookup tables to gain respectably quick computation.

The first step is to construct a sine-cosine lookup table. A cycle length of 256 entries has been used here. The one-dimensional Fourier transform of a spoke

FIGURE 11. Backprojection contributions.

from the sampling lattice of the Hermitianized signal $H(R, \theta \pm \pi)$ will serve as the secondary lookup table. Although an FFT can be used to construct that one-dimensional transform, since there are so few samples and the output is purely real there is little advantage over classical synthesis.

The next step is to add, by projection, the results from each spoke into an output image raster. A couple of examples for a double star are shown in Figure 11, where a pair of sinc functions get projected onto the raster. When all of the projection angles are included, the result is the image shown in Figure 12. In this case, the point spread function is quite reasonable because the $1/r$ density distribution of the sampling lattice shown in Figure 10 is not too far from the roughly conical form that would be the autocorrelation of a circular aperture. As such, this unfiltered backprojection is suitable. When there are more radii involved, however, it is preferable to emphasize the outer ones to achieve a better point spread function. Such a procedure is called *filtered backprojection* and is popular in medical imaging.

## Monte Carlo Simulation

The effects of quantum noise can be simulated realistically with a Monte Carlo approach where the $T_{\pm}(R, \theta)$ values are treated as relative probabilities of transmission. First we set up a companion array of the same dimensions that is entirely initialized to zero. Then we figuratively start throwing photons at the collimators, with each photon being characterized by two random numbers. The first of the numbers is uniform between 1 and the number of $T$ samples (1152 in our particular simulation here). If the second random number, uniform from 0 to 1, is

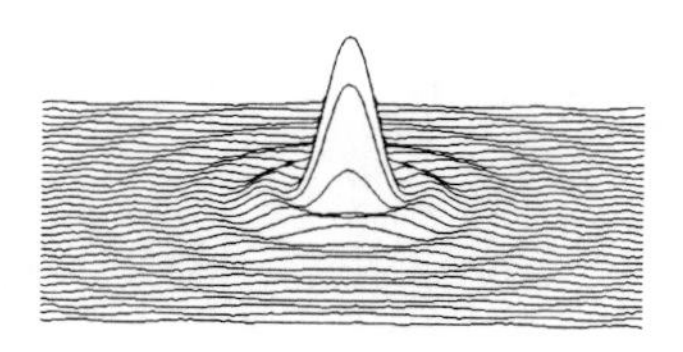

FIGURE 12. Point spread function.

lower than the normalized $T$ value, then we increment the corresponding value of the companion array to count it as a detected photon. On the other hand, if the second random number exceeds the normalized $T$ value, we presume that the photon failed to get all the way through the collimator. This Monte Carlo process is continued until a preset total number of detected photons is tallied.

The companion array now may be treated, just as if it were $T$ values, with the Hermitianizing recipe and Fourier synthesis. Figure 13 shows the results for an image composed of three stars. The peaks have been normalized to the same height for clarity in the illustrations, although the program first provides heights proportional to exposure. For reference, the upper-left-hand panel of Figure 13 shows the image reconstructed directly from the $T$ samples, which is tantamount to having an infinite number of photons.

Things to notice are side lobes and statistical noise. The side lobes appear here as small ripples. The side lobe reduction that has come from using many collimator pairs is essential. Nevertheless, at a distance equal to the width of the field there is a strong ring side lobe that only appears when there is a strong source near one of the corners of the field. Along with the vignetting of the collimator tubes, that ring side lobe limits the extent of the useful field. The bump in the lower-left corners of Figure 13 is just such a side lobe of the main source shown.

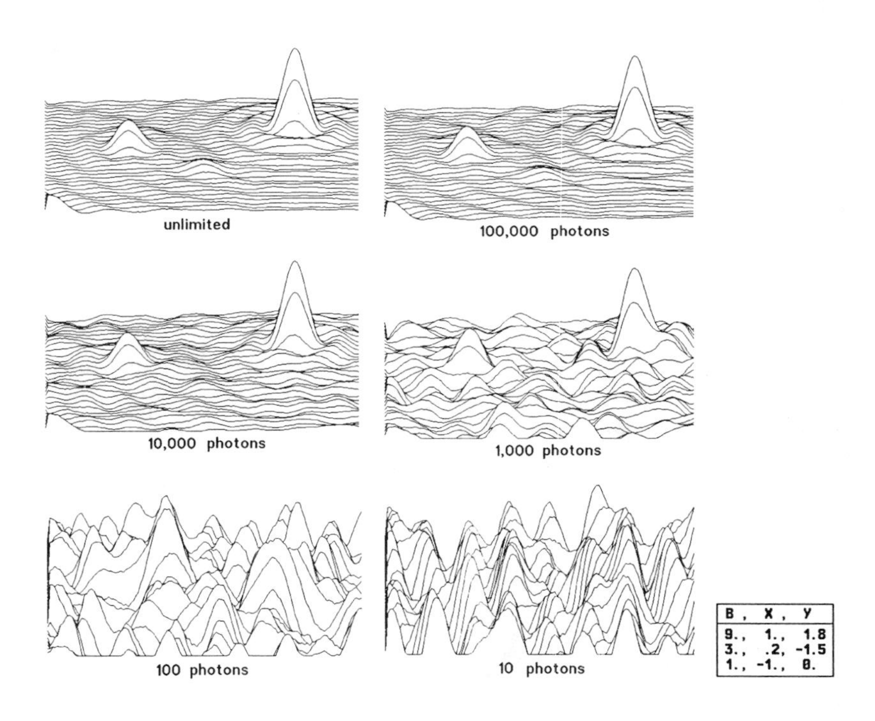


FIGURE 13. Monte Carlo results for an image of three stars.

The statistical noise appears as choppiness. This increases according to the square root of the total number of photons; but since the signal increases directly with the number of photons, there is a clear net increase in signal-to-noise with more photons. In comparison with direct imaging, where the photon statistical noise stays localized in the images, the noise power is distributed uniformly over the scene. For the same number of photons, the total noise power is the same for both situations. As a result, the indirect imaging results in a net signal-to-noise gain for synthetic imaging for those portions of the image that are above the rms illumination, but a degradation for the background that lies below the rms level. That degradation has been a main objection to synthetic imaging, because for astronomers it is usually more important to see very faint sources in the same field as bright sources than to have accurate photometry of the bright sources. Such objection is not at all justified, however, since the discouraging noise assessment is based completely on linear image reconstruction.

Over the past 25 years there have been developed a wide variety of modern spectral estimates that can be vastly superior to conventional Fourier analysis. One of the most celebrated of these has been maximum-entropy spectral analysis [Burg 1967]. The important recognition was that, rather than assume the information beyond the range of measurement to be zero, a Bayesian approach should be adopted so as to find the smoothest reconstruction that would agree with the measured data. Ables [1974] summed it up very nicely with his dictum "The result of any transformation imposed on the experimental data shall incorporate and be consistent with all the relevant data and be maximally non-committal with regard to unavailable data." These modern spectral estimates are implicitly constrained to be nonnegative, and their success sometimes is attributed to the valid extra information incorporated by that constraint. There also would be a certain rationale in adopting a minimum-entropy criterion [Wiggins 1978; Ulrych & Walker 1982] whereby the picture is presumed to be composed of a discrete number of unresolvable stars, as opposed to maximum entropy that presumes the smoothest picture.

While there are no analytic expressions for the signal-to-noise behavior of these nonlinear methods, there have been a few empirical trials. Hanson [1986] and Murphy [1990] have shown that the background is flattened away while legitimate weak sources are preserved and that no artifacts are introduced. A further feature of Murphy's result was considerably sharper resolution than was obtained from Fourier processing. The important deduction is that their results nullify the main objection to indirect-imaging X-ray telescopes. Both authors used their own two-dimensional coding, and Minerbo [1979] has described a maximum-entropy routine that is specifically adapted to the polar sampling lattice although its noise behavior has not been tested in the same context.

A somewhat simpler approach might be obtained from a blend of one-dimensional algorithms. The idea would be to perform modern spectral estimates of the information on the spokes of the sampling lattice and use those estimates rather than the conventional Fourier expressions for the lookup backprojection functions comparable to those of Figure 11. Modern spectral estimates typically show Lorentzian $\exp(-|x|)$ as opposed to $\text{sinc}(x)$ point spread functions, even though the convolutional concept of point spread functions is no longer strictly applicable. The Lorentzians are much sharper and have lower as well as non-negative skirts. Figure 14 shows roughly what would get used in place of Figure 11, and it is not hard to imagine the improvements that would accrue. The main benefit would be a flattening of the background so as not to conceal legitimate faint sources, but the improved resolution is also welcome. On the other hand, that extra sharpness may call for more radial spokes in the sampling lattice to avoid the appearance of radially oriented side lobes.

There are numerous published algorithms for the one-dimensional situation. On my agenda for things to be done in the future is to try some of those published in the books by Marple [1987] and Kay [1987] in this possible backprojection application. Even if they turn out not to perform well, the fully two-dimensional codes still serve as a fallback.

The fact that astronomical images usually are composed of discrete stellar point sources suggests that a minimum-entropy solution may be more appropriate than maximum entropy, because the latter gives the smoothest solution. Minimum entropy would be in keeping with the previously mentioned quest for an exact solution to a single pair of collimators using a minimum discrete number of stars. Wiggins [1978] published one-dimensional minimum entropy techniques for geophysical data analysis.

## Upscale Rendering

Rotational aperture synthesis has proven to be extremely successful in radio astronomy. The images generated show much more detail with much more dynamic range than direct optical telescopic images, and those qualities have been a clear result of nonlinear processing strategies that give improvements over raw Fourier analysis. It is not unreasonable to expect comparable success for rotational aperture synthesis at X-ray wavelengths. Moreover, the X-ray regime should be the easiest regime to achieve aperture synthesis for the following reasons. Unlike the radio or optical fringe responses, those for the modulation collimator are achromatic, so that broad spectral bands can be used to gain information from faint sources.

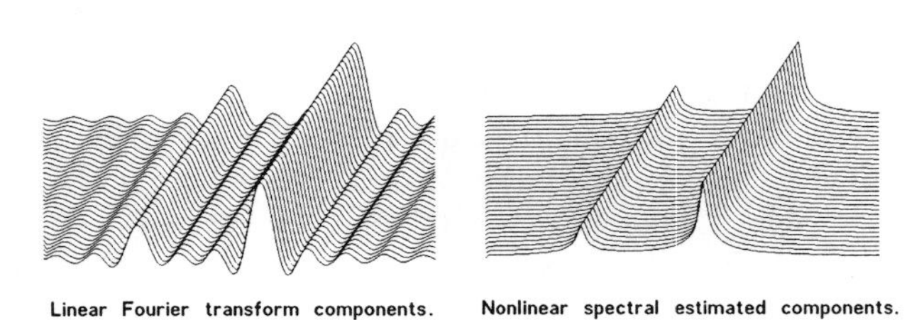


FIGURE 14. Backprojection contributions from a spoke of data for different estimators.

Admittedly, we are not achieving interferometric resolution, but we are achieving resolution that is at least as good and probably better than that available by direct imaging. It is also easier to achieve a pure regular polar sampling lattice with a desired rotational axis than if we have to depend on earth rotation. This control over the sampling lattice also allows wider selection of field and resolution trade-offs.

Figure 15 shows the kind of X-ray telescope that I have in mind. It would be composed of several hundred collimators, each several centimeters in diameter. The reason for so many is to gain a large number of resolved pixels in the field of view. Kilner found by empirical simulations such as those appearing in Figure 13 that there is no signal-to-noise penalty from dividing up a given aperture area into more and more smaller collimators for achieving higher resolution. The benefit purely is more and more resolution within the field. Furthermore, it is much easier and less expensive to fabricate many detectors of a few centimeters size to fill the aperture than only a few much larger detectors. Silicon detectors are particularly suitable candidates for operation up to about 60 KeV X-rays. To be sure, there are more electronics, but most of it is duplicated circuitry. The vehicle would rotate at several revolutions per minute. That should be fast enough for stabilization and

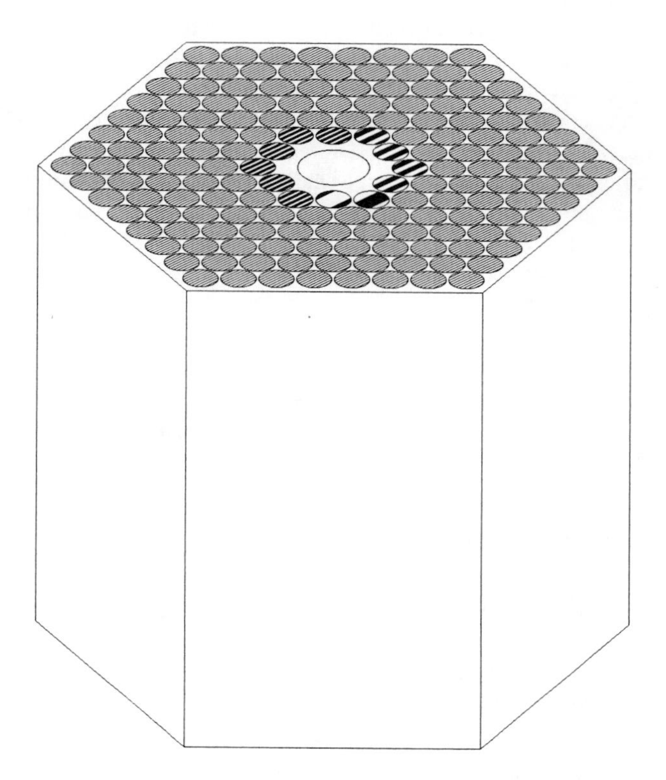

FIGURE 15. Commendable synthetic-aperture x-ray telescope.

not too fast for reorientation to view different directions. It should also be fast enough to reasonably cope with transient burst sources. Even if a full rotation is not available during a burst, there would be considerable positional information available. Somewhat more positional information could result if the collimators were not all disposed with their bars parallel, although that disposition might somewhat complicate $\delta$ corrections for slight misalignment of the spin axis with respect to projected bar patterns.

In comparison with the grazing-incidence telescopes that have become so popular, there would be many merits for rotational aperture synthesis. First and most important is that the sensitivity extends much farther to reach the hard X-ray regime. The hard limit is only when the bar grids cannot be thick enough to obstruct the rays. It should be most important to cover this range and establish a well-resolved relation between X-ray sources and gamma-ray sources. Economics and efficiency give further merits. Grazing-incidence reflectors are very heavy and extremely expensive to fabricate at the high precision required. Because they are grazing, it is also necessary to fabricate surface areas that are hundreds of times more than the actual collection area. The collimator grids would be small and light and give 25% area efficiency.

Grazing incidence requires position-sensitive detectors and three-axis vehicle stabilization. For aperture synthesis the detectors can be chosen for high quantum efficiency with good color discrimination, and the vehicle is spin-stabilized. In fairness, the counter argument is that the position-sensitive detector will have less background contamination because each pixel is smaller than each individual synthesis detector. The importance of that factor cannot be properly assessed until there is more experience with the nonlinear reconstruction capabilities, but it is hard to imagine that it would be sufficient to outweigh all of the advantages of a synthesis telescope.

## Bibliography

J. G. Ables, 1968 Proc. Astron. Soc. Aust. 4: 172

———1974 "Maximum entropy spectral analysis" Astron. and Astrophys. Suppl. Ser. 15: 383–393

R. N. Bracewell, 1986 *The Hartley Transform,* Clarendon Press

J. P. Burg, 1967 "Maximum entropy spectral analysis" *Proc. 37th Meeting of the Society of Exploration Geophysicists.*

R. H. Dicke, 1968 "Scatter-hole cameras for x-rays and gamma rays" *Astrophys. J.* 153: L101

K. M. Hanson, 1986 "Effects of nonnegativity constraints on detectability" in *Quantum-limited Imaging and Image Processing,* Optical Society of America, Technical Digest

S. M. Kay, 1987 *Modern Spectral Estimation,* Prentice Hall

A. J. Levy, 1983 "A fast quadratic programming algorithm for positive signal restoration" *IEEE Trans.* ASSP-31: 1337

S. L. Marple, 1987 *Digital Spectral Analysis,* Prentice Hall

L. N. Mertz, 1967 "A dilute image transform with application to an x-ray star camera" in *Modern Optics,* ed. J. Fox, vol.17, Polytechnic Inst. of Brooklyn

———1976 "Positively constrained imagery for rotation collimators" *Astrophys. Space Sci.* 45: 383–389

———1989 "Ancestry of indirect techniques for x-ray imaging" *Proc. SPIE* 1195: 14– 17

L. N. Mertz, G. H. Nakano and J. R. Kilner, 1986 "Rotational aperture synthesis for x-rays" *J. Opt. Soc. Am. A* 3: 2167–2170

L. Mertz and N. O. Young, 1962 "Fresnel transformations of images" in *Optical Instruments and Techniques,* Chapman and Hall

G. Minerbo, 1979 "MENT: A maximum entropy algorithm for reconstructing a source from projection data" *Comp. Graphics and Image Proc.* 10: 48–68

M. J. Murphy, 1990 "The virtues of positive-definite reconstruction of x-ray and gamma-ray images" *Nucl. Inst. and Methods* A290: 551–558

M. Oda, 1965 "High-resolution x-ray collimator with broad field of view for astronomical use" *Appl. Opt.* 4: 143

G. L. Rogers, 1977 *Noncoherent Optical Processing,* Wiley

H. W. Schnopper *et al.,* 1970 "Precise location of Saggitarius x-ray sources with a rocket-borne rotating modulation collimator" *Astrophys. J.* 161: L161–167

T. J. Ulrych and C. Walker, 1982 "Analytic minimum entropy deconvolution" *Geophysics* 47: 1295–1302

E. T. Whittaker and G. N. Watson, 1915 *A Course in Modern Analysis,* Cambridge University Press

R. A. Wiggins, 1978 "Minimum entropy deconvolution" *Geoexploration* 16: 21–35

N. O. Young, 1963 *Sky and Telescope* 25: 8–9

# 3

# Interferometry

## Introduction

The cornerstone of this chapter is the pragmatic assertion that phase and position are identically one and the same variable. Perhaps it is presumptuous, but nevertheless I do like to think of the assertion as a lesser scale analog of Keats' identification of beauty with truth. My intent is to show the plausibility of the assertion in connection with a variety of experiments, and to indicate its practical and philosophical implications. All too frequently in the past we had been led to believe that we could not measure phase in the optical region as contrasted with the radio region. An equivalence of phase with position makes the measurement not only possible but commonplace.

Start from Young's well-known experiment shown in Figure 1. Certainly in this instance phase corresponds to the position in the plane of the fringes. It is perfectly easy then with a position-sensitive photon detector to measure the phase of an individual photon. The fringe pattern develops from an ensemble of photons with their phase probabilities. Before delving further into Young's experiment, however, it will be worthwhile to review customary interferometric techniques for comparison.

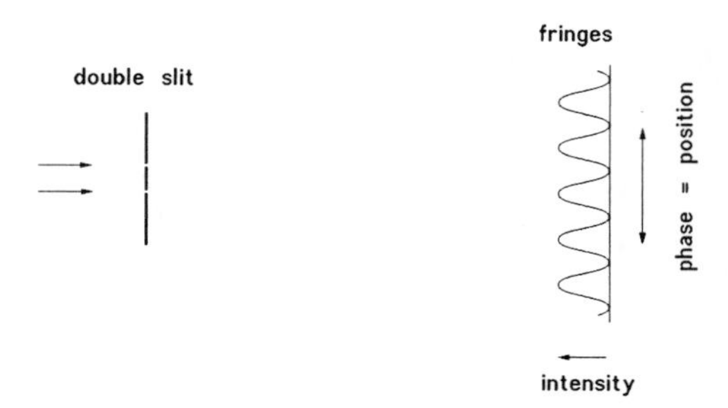


FIGURE 1. Young's double-slit experiment.

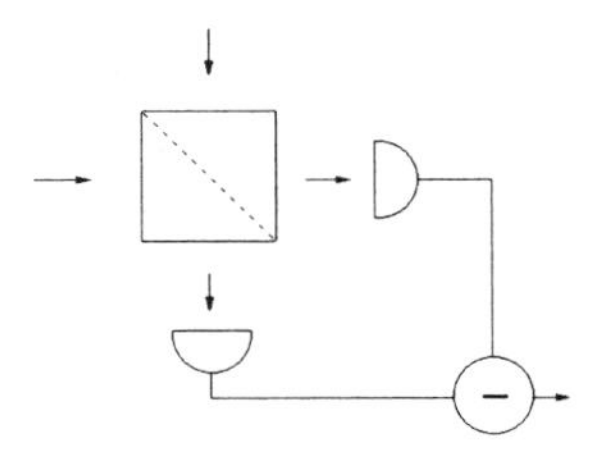


FIGURE 2. Balanced homodyne mixing.

The usual way to combine or mix two beams of light is with a semitransparent beamsplitter as shown in Figure 2. There are two inputs and two outputs. When the difference of the two outputs is measured from two detectors at the outputs, the configuration often is called a *balanced homodyne mixer*. It turns out that the balancing only works for dielectric beamsplitters, where the absence of absorption implies that the outputs must be complementary. It is most instructive to perform the experiment shown in Figures 3 and 4. Sagnac's cyclic interferometer is particularly easy to to adjust, even with white light, by aligning the two images seen through it. Arrange it with illuminated white cards so that the fringes by transmission appear juxtaposed with those seen by reflection. As expected, for a dielectric beamsplitter the reflection fringes are complementary or 180° out of phase with the transmission fringes, as is depicted in Figure 4A. For a metallic beamsplitter, however, the fringes are observed to be precisely in phase, as shown in Figure 4B. One way to look at the situation is that the two beams impinging on the beamsplitter create a field of standing waves. The absorption by the metallic film is greatly changed depending on whether the film lies in a node or a crest of those standing waves. When it lies in a crest, the absorption is total because both outputs are completely black.

It might be quite interesting if one could make a combination dielectric-metallic beamsplitter such that the outputs would emerge in quadrature, with sinusoidal

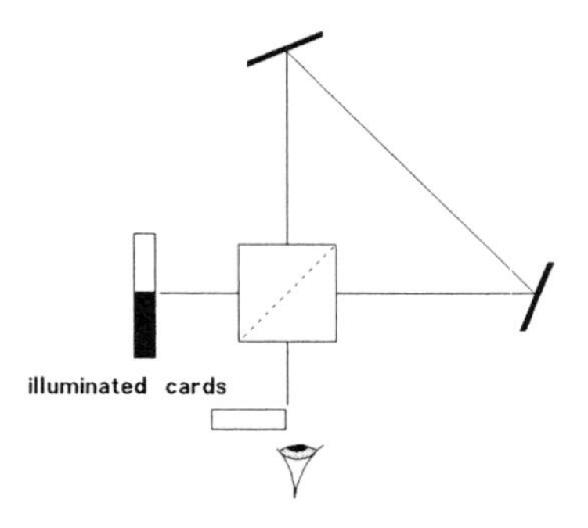


FIGURE 3. Sagnac's cyclic interferometer.

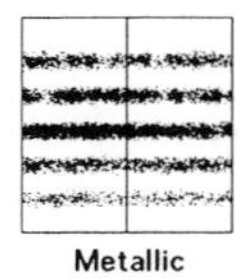


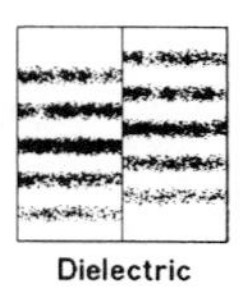


FIGURE 4. Fringe appearance with A: metallic beamsplitter, B: dielectric beamsplitter.

versus cosinusoidal fringes. That would confer a sense of directionality on the fringes to distinguish increasing from decreasing interferometric path differences. At present I am unaware of any such recipe, but in any case it has become somewhat academic. In the first place, it still would be necessary to have a third detector someplace to monitor the total intensity, and in the second place we now have available efficient and simple homodyne mixers with three or more ports, as will be described shortly.

The balance of the balanced homodyne mixer shown in Figure 2 only applies when the output beams have equal intensity, so that common-mode intensity fluctuations cancel out in the differential signal from the detectors. Complete immunity from intensity fluctuations can be had by measuring the fringe visibility, which is the ratio of the difference to the sum signal from the two detectors. In a photon-counting mode ratiometric finges can be obtained by having each count arriving at one detector set a flip-flop while each arriving at the other detector clears the flip-flop. The average value of the flip-flop output provides the ratiometric signal. The statistics of the result are such that if either detector sees a truly dark fringe, then there will be no chatter at all of the flip-flop. Maximum chatter occurs when the individual signals are balanced. Notice that this susceptibility is the converse of the cancellation for large-signal intensity fluctuations in the nonratiometric operation. Both situations give two peaks of noise power per fringe cycle, but the peaks occur at different phases in that cycle. Furthermore, it should be noticed that the photon-counting ratiometric mode implicitly depends on noncorrelation of the photons. For example, if the photon detections are rare and in equal numbers, but with one channel systematically receiving a photon just after the other channel, then it will appear that the later channel is always on.

## Triphase Reception

The fundamental problem with the beamsplitter or balanced homodyne mixer as shown in Figure 2 is that it has only two outputs, so it provides insufficient information to assess independently the brightness, contrast, and phase of the fringes. At least one more port is essential. The extra port or ports can be developed from Figure 1. All that is necessary is to slice the Young's fringe pattern, like bread, into three or more slices per fringe. One way to do the slicing is as shown in Figure 5, with a lenticular screen, which is just a thin sheet of transparent plastic with a

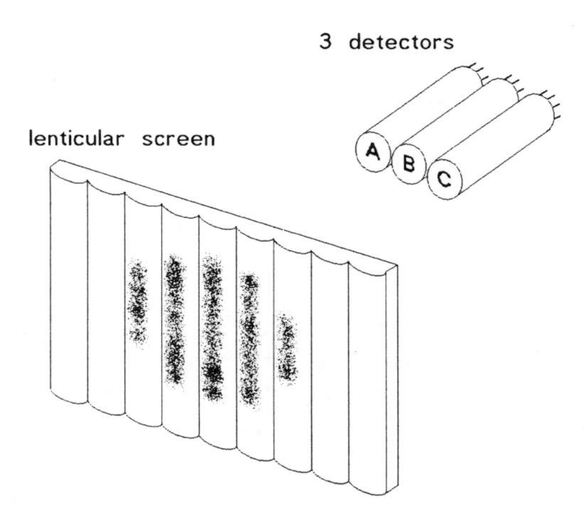


FIGURE 5. Lenticular receiver.

corduroy-like texture. The idea is to match the spacing of the fringes, by adjusting the angle between the incident beams, to that of the cylindrical lenticles on the screen. Light falling on the crests of the lenticles goes straight through, while light falling on the flanks gets diverted to present a fan beam. The screen should be fine, but not too fine; 40 lenticles per millimeter works quite well. Three or more discrete detectors divide the fan beam, which actually is composed of numerous diffraction orders, into uniform segments that serve as output ports. For a three-port mixer, the signals have the form

$$A = \alpha + \beta \cos(\phi - 120°) \ ,$$
$$B = \alpha + \beta \cos(\phi) \ ,$$
$$C = \alpha + \beta \cos(\phi + 120°) \ .$$

Three is a necessary and sufficient number of ports to solve independently for $\alpha$, $\beta$, and $\phi$. For fixed $\alpha$ and $\beta$ these three signals are shown as a function of phase $\phi$ in Figure 6.

A very simple interferometer for exercising the mixer is shown in Figure 7. The dihedral arrangement with a pivot at its vertex allows $\phi$ to be scanned conveniently

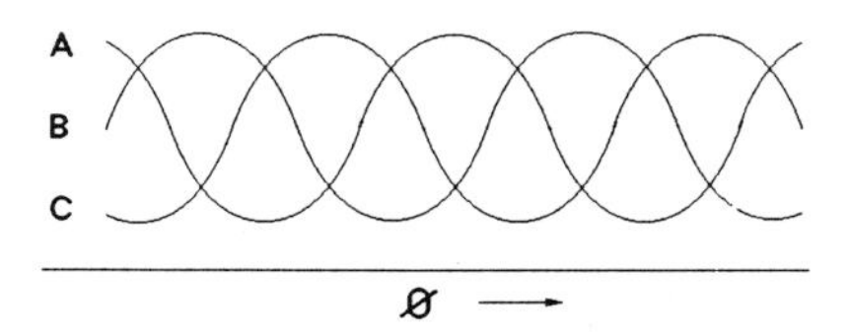


FIGURE 6. Triphase signals.

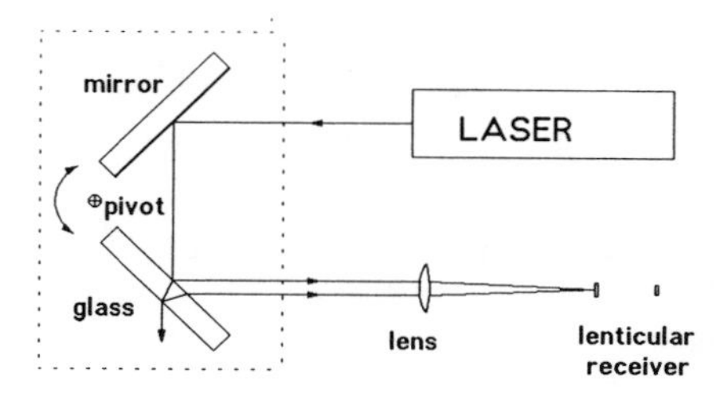


FIGURE 7. Convenient interferometer.

without upsetting the beam alignment. The choice of a green laser facilitates the use of photomultipliers having bialkali photocathodes for photon counting. Drastic attenuation of the beam keeps the count rate reasonable. A particularly convenient circuit for the counting is shown in Figure 8. The 74LS63 is a hex current-sensing buffer and is biased just above threshold current with the resistors. Charge pulses from the photomultiplier anode pull down the current to result in clean 20-nanosecond TTL pulses as shown. I am particularly fond of the circuit because of its simplicity and my frugality; six channels of preamplifier, discriminator, and pulse-shaper come on a chip that costs less than a dollar.

The compact circuitry of Figure 9 provides ratiometry and display of the counts in the Argand diagram. The oscilloscope spot goes toward one vertex of the display triangle, depending on which photomultiplier most recently sensed a count. When the count rate is low and the capacitors are small or absent, the spot has time to arrive at the vertex as shown in the time exposure of the left oscilloscope trace. When the count rate is higher and the capacitors are included, then the spot jiggles around some mean position within the triangle. As the interferometer is scanned the jiggling spot traces around a roughly elliptical (ideally circular) path as in the right trace, clockwise or counterclockwise, depending on the scanning direction. The radial size of the elliptical path is proportional to the fringe contrast of the interferometer. As long as that size is sufficient for the trajectory to stay clear of the center of the triangle, it should be possible to unwrap the phase and count fringes without error.

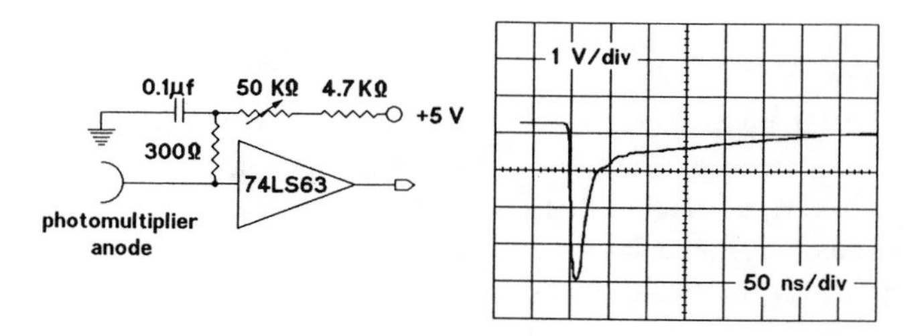


FIGURE 8. Photon-counting circuit.

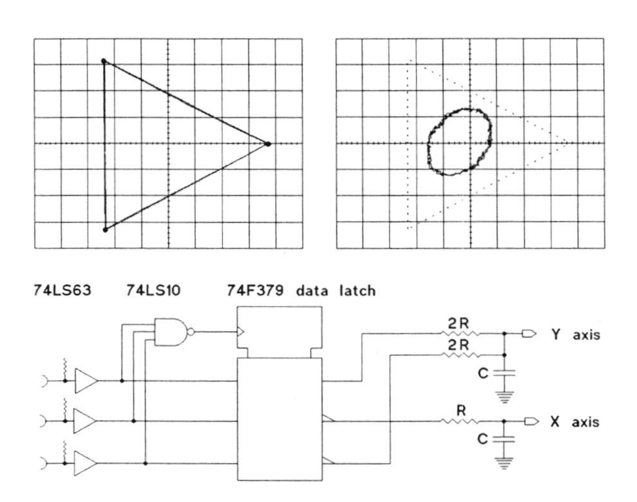


FIGURE 9. Ratiometric oscilloscope Argand displays with circuit.

## Unwrapping and Filtering

Phase unwrapping is achieved with the recursive procedure

$$\Phi_{t+1} = \Phi_t - (\Phi_t - \phi_t)_R \ ,$$

where $\Phi$ is the unwrapped phase that includes the proper fringe count, $\phi$ is the phase measurement input (which in this photon- counting case can only take on the values 0 or $\pm\frac{1}{3}$ cycle), and the subscript $R$ signifies that the parenthetical expression is reduced to the interval $\pm$ a half-cycle. That reduction, which is the same as is always done for evaluating trigonometric functions, ensures that successive phases $\Phi$ differ by less than half a cycle. The fractional part of the phase $\Phi_t$ cancels out so that the fractional part of $\Phi_{t+1}$ agrees with that of $\phi_t$. The unit difference of the temporal index denotes the latency or delay of the output.

The structure of that recursive relation is much the same as that of a single-pole recursive low-pass digital filter. The inclusion of a decay constant $N$ increases that similarity:

$$\Phi_{t+1} = \Phi_t - N^{-1}(\Phi_t - \phi_t)_R \ .$$

Now the recursion gives an exponentially weighted running average of effectively the prior $N$ samples of $\phi$. The subscript $R$ is accomplished by throwing away the overflow bit of the subtraction that is in the parentheses, and so actually utilizes what always has been known as the overflow instability of such a digital filter. With some averaging, $N > 1$, the phase unwrapping turns out to be more tenacious, with fewer errors, wherein the output skips or slips cycles. On the other hand, there is also a diminished maximum slew rate with which the recursion can keep up.

This recursive unwrapping filter is readily implemented with a PROM (Programable Read-Only Memory), adder, and Multiplier-Accumulator as shown in Figure 10. The MAC should be wired for two's complement operation with no rounding. Since only three words (A, B, or C) need be selected, it is even easier to lump the adder as a look-up table in with the PROM, but the operation is clearer when shown separately. The decay constant $N$ may be entered into the MAC with switches. A curious aspect of this filter is that the decay constant is elastic with time. The clock $t$ for the MAC gets activated by the photon counts themselves, and the decay time becomes elastic. Lower photon rates extend the decay time so as to always encompass the equivalent of exactly $N \pm 0$ photons within the filter. The resulting statistics are quite different from those derived from a Poisson process where there are $N \pm \sqrt{N}$ photons encompassed in a fixed time decay.

This recursive filter contrasts with the usual procedures for phase estimation, which generally do the averaging process on the sine and cosine quadrature components of the phase and then evaluate the arctangent. It is quite feasible to set up similar recursive relations for those quadrature components:

$$S_{t+1} = S_t - N^{-1}(S_t - \sin \phi_t) \ ,$$

$$C_{t+1} = C_t - N^{-1}(C_t - \cos \phi_t) \ ,$$

and then transform to polar coordinates:

$$\Phi_{t+1} = \text{nint}(\Phi_t) + \frac{1}{2\pi} \arctan \frac{S_{t+1}}{C_{t+1}} \ ,$$

where nint is the nearest integer function to take care of the phase unwrapping, and

$$\Gamma_{t+1} = \sqrt{S_{t+1}^2 + C_{t+1}^2} \ ,$$

where $\Gamma$ is the fringe contrast. In practice, there are awkward problems with these relations. First, for a balanced homodyne receiver there are only two ports, so the phase contribution from each photon takes on only two possible values, $\pm 90°$, and that would lead to the sine recursion filter. A shift of reference point can be introduced subsequently by changing the path difference of the interferometer to assign the two values as 0° and 180°, and thus gain the cosine recursion filter; but both values are not available simultaneously. That problem could be solved by using the lenticular scheme as a four-port homodyne receiver, although that

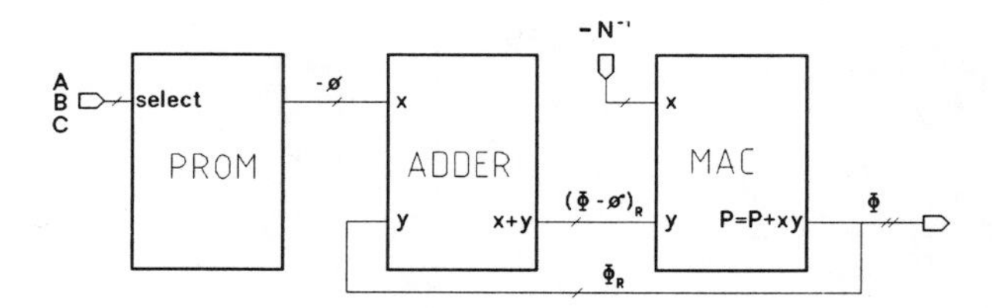


FIGURE 10. Digital phase unwrapping and low-pass filter.

solution has never been tried. A second problem is that phase unwrapping errors occur more readily with these cartesian recursion relations than with the more direct polar recursion. Third, the polar recursion relation requires far less computational effort.

If desired, the phase recursion relation can be accompanied by an associated contrast recursion of the form,

$$\Gamma_{t+1} = \Gamma_t - N^{-1}(\Gamma_t - \cos(\Phi_t - \phi_t)) \ .$$

For most metrology applications, however, the contrast $\Gamma$ is superfluous. Another associated recursion might be for the error,

$$E_{t+1} = E_t - N^{-1}(E_t - (\Phi_t - \phi_t)^2_R) \ ,$$

where the general idea would be to minimize the error $E$ by judicious adjustment of the decay constant $N$. This is not very different from maximizing the contrast $\Gamma$, since we may note that the cosine function exhibits a maximum at $(\Phi - \phi) = 0$, whereas the parabolic error exhibits a minimum there. For example, if we wish to photograph a jiggling optical fringe pattern, we will clearly obtain the maximum photographic fringe contrast with the minimum camera-tracking error of the optical fringe phase. Thus we would like to choose $N$ so as to maximize $\Gamma$. This choice is most easily done by trial and error if the original phase $\phi$ data are fully recorded; otherwise, a Kalman filter optimization approach might be devised.

In the original report introducing the phase recursion relation there were two figures, both reproduced here as Figure 11. These show tracking simulations of a sinusoidally wandering low-contrast fringe pattern. On the left is the simulation for the polar recursion and the one on the right, with the same data, is for the cartesian recursions. The bar length in between shows one fringe spacing of the pattern being tracked. Clearly the cartesian recursions make frequent phase unwrapping errors. The circumstances of course were chosen to best emphasize the merits of polar recursion.

In attempting to assay the performance more quantitatively, things to notice are that certain parts of the filtering are nonlinear and so an empirical Monte

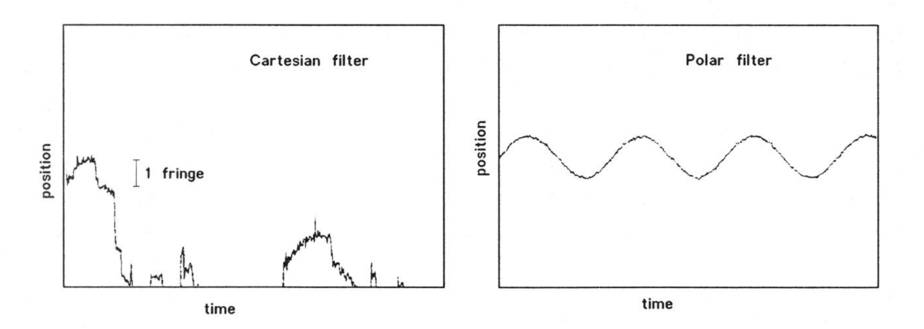


FIGURE 11. Comparative performance of cartesian versus polar tracking filter.

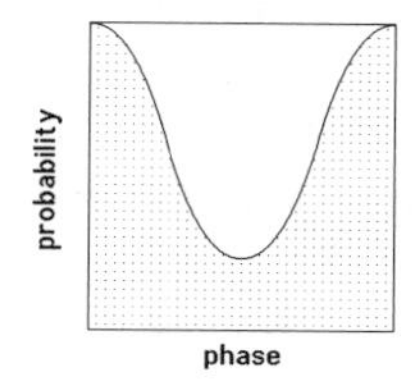


FIGURE 12. Probability lookup table for Monte Carlo simulation.

Carlo approach should be easier than an analytic one. There are not too many free parameters. The main ones are the fringe contrast and the ratio of the fringe slew rate to the delay constant $N$. A more subtle free parameter is the number of bins used to characterize the phase $\phi$. The first step toward constructing a Monte Carlo simulation of moving fringes of a selected contrast is to make a sine-wave lookup table as shown in Figure 12. The ratio of the peak minus the minimum to the peak minus zero is our fringe contrast. Two random bytes specify coordinates for a point in the figure. If the point lies above the curve, we reject it and try two more random bytes. If the point lies in the shaded region below the curve, we add a systematic slew value to the abscissa to get the phase $\phi$ for that accepted photon. That phase $\phi$ then can be rounded to the nearest bin by which we are defining phase. Three bins are necessary and sufficient for the filter operation, but more can be used up to the full 256 specified by the byte. The filter output $\Phi$ for the onslaught of photons can be displayed on the screen as shown in Figure 13. This is rather like Figure 11 except for linear rather than sinusoidal slewing. The vertical scale range shown is four fringes, and occasional phase unwrapping errors are conspicuous.

Counting the errors, or lack thereof, to see how often they occur for various parameter choices shows the statistical limits of performance. For example, it takes about 10 photons ($N = 10$) to define the phase of full-contrast fringes ($\beta = \alpha$) for reliable phase unwrapping. At the other extreme, the slew rate should be significantly less than a half-fringe per decay constant $N$ for reliable tracking.

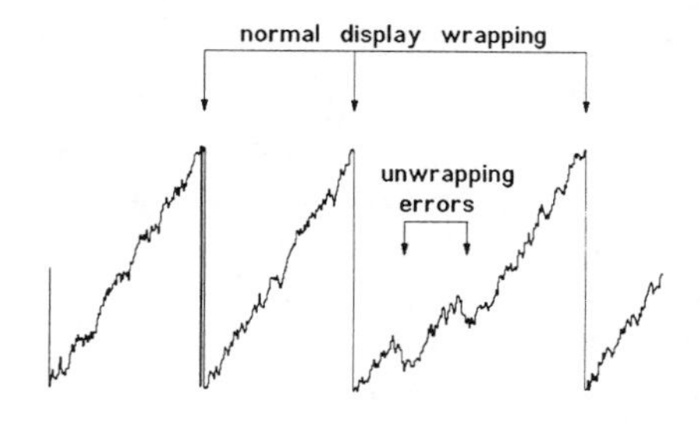


FIGURE 13. Phase unwrapping errors. Full scale is four fringes; two errors are apparent.

That tracking also degrades substantially for only a few phase bins, such as the minimum of three.

The very fact that we can easily identify phase unwrapping errors in Figure 13 indicates that the phase filter is still not optimal. An optimal filter also should be able to recognize and then correct those errors.

In the absence of any phase unwrapping errors, the mean square width of the cosine bell fringe probability distribution function is given by

$$\frac{\int_0^\pi \phi^2 P(\phi)\, d\phi}{\int_0^\pi P(\phi)\, d\phi} = \frac{\int_0^\pi \phi^2 (1+\beta\cos\phi)\, d\phi}{\int_0^\pi (1+\beta\cos\phi)\, d\phi} \quad .$$

Evaluating the expression gives $(\pi^2 - 6\beta)/3$, whose square root gives 1.14 rad as the root mean square error for $\beta = 1$ unity contrast fringes. That is about 0.18 fringe rms error per photon. For an ensemble of photons depicting a stationary fringe pattern, that shot noise uncertainty should decrease in proportion to the square root of the number of photons in the ensemble, but remember that this estimate is contingent on no unwrapping errors at all.

## Substantial Intensities

Almost all light sources, especially lasers, are far too bright to be able to count photons. Hence there is a motivation to adapt the three-port mixing for analog signals. The first step is to replace the photomultiplier detectors of Figure 6 with silicon photodetectors to gain simplicity and quantum efficiency. An even better approach that has been invented and implemented at Optra, Inc. (Topsfield, MA), is to replace the lenticular screen with an array of silicon photodetectors where every third element is connected together internally. In effect there are three detectors having triply interlaced areas on a single substrate. There are major advantages to this approach in that the three resulting signals are more assuredly matched with $\pm 120^\circ$ phase differences and without requiring careful alignment. This triphase detector is also less affected by differential intensity fluctuations of the interfering light beams.

Linear combinations of the three $120^\circ$ signals lead to,

$$\begin{aligned}
\beta \sin\phi &= (+A \qquad\qquad -C)/\sqrt{3}\ , \\
\beta \cos\phi &= (-A \quad +2B \quad -C)/\,3\ , \\
\alpha &= (+A \quad +B \quad +C)/\,3\ .
\end{aligned}$$

The signs of the first two equations give the Gray code representation of $\phi \pm 180^\circ$, resolving $\phi$ to $90^\circ$ and allowing easy bidirectional fringe counting. The circuitry of Figure 14 digitizes these two equations and counts fringes. The unwrapping is a simpler version of that done back with Figure 10 and proves to be particularly adept for comparing two separate lasers that impinge on the lenticular screen shown back in Figure 5, because it not only gives a difference frequency like that from ordinary balanced homodyne mixers, but it also gives the sign of that difference frequency. Thus there is no need to introduce any frequency offset, such as from an

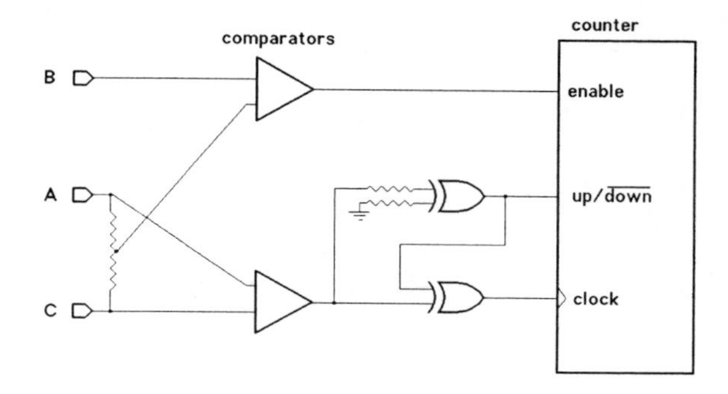


FIGURE 14. Fringe counting circuit.

acousto-optic light modulator, for the frequency comparison. More experimental details about such comparisons appear later in this chapter.

## Flash Conversion

The resistive divider of Figure 14 can be omitted by comparing all three pairs of the $ABC$ inputs and decoding the results to improve the resolution to 60°. Subsequently a much bigger resolution improvement can be gained by interpolating within each 60° sector with a flash analog-to-digital (A-D) converter. I call the circuit of Figure 15 a *hexaflash circuit* because it uses six flash A-D converters to do just that. A flash, or parallel, A-D converter contains a voltage divider with numerous taps. The input signal voltage is compared simultaneously with the voltage at each tap, and the result is encoded in binary format. Ordinarily, the high end of the voltage divider is connected to a fixed reference voltage and the low end to ground, but variable voltages are acceptable in practice. The only stipulations are that the high end be at a higher voltage than the low end, and that the signal be in between the two reference voltages. Well, there are only six permutations of the three-phase input signals, so these are hooked up to six converters. Buffer amplifiers are important to fulfill the heavy loading imposed by all of the converters. The converters that I use also have an overflow bit that compares the signal with the high reference. Three of these overflow comparisons get decoded to enable the output of only the valid converter.

Referring back to Figure 6, the upper and lower envelopes of the signals apply to the high and low reference voltages, while the triangular-wave intermediate value applies to the input signal for the valid converter. The output of the valid converter is concatenated with the three comparison bits to address a PROM that has been programmed to yield the output phase in binary format. For 6-bit converters the phase resolution is 1/(6 x 64) or 1/384th of a cycle, but because most electronics deal with byte-wide numbers, the result may be rounded to 1/256th of a cycle. For

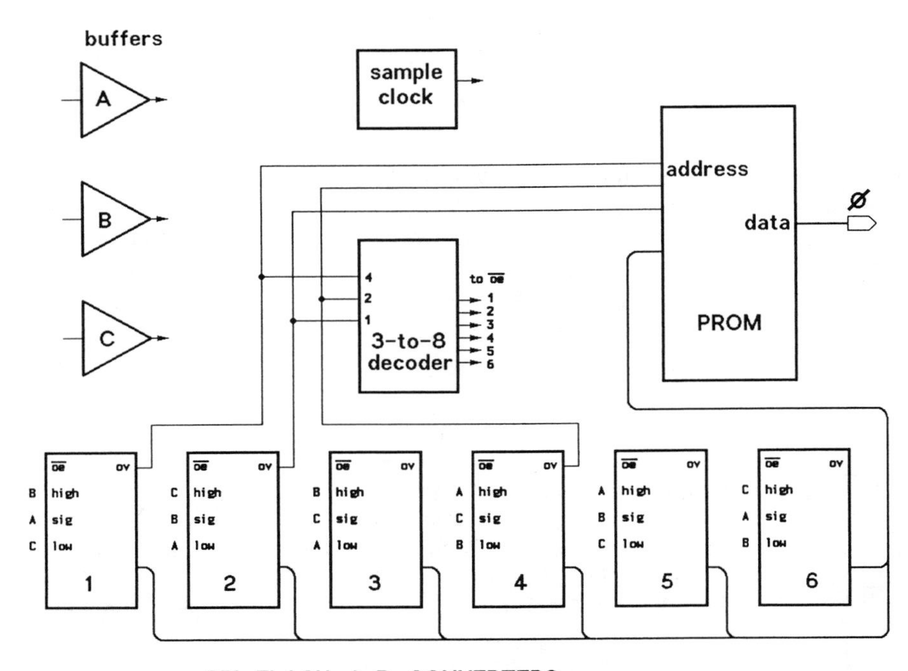


FIGURE 15. Hexaflash three-phase to digital converter.

8-bit converters, the result is good to 1/1536th of a cycle, and can be approximately represented with 12- or 16-bit numbers.

A very important merit of the combined three-phase mixer and hexaflash circuit is that it is completely static, so that it does not need any modulators and can operate at high bandwidths. The requisite phase shifts have been organized spatially rather than temporally. Furthermore, the scheme works equally well at any operating point of the fringes. There is no need to adjust an optical delay to seek a balance point.

The hardware implementation of the phase filter was shown back in Figure 10. To accommodate 1536 phase values, the words should be at least 12 bits wide. Either 12- or 16-bit MACs are readily available, but the 16-bit version has one input multiplexed with the output to reduce the pin count, and so is less convenient to use in a stand-alone circuit. The 12-bit version has a 27-bit accumulator, which can be extended much like counting fringes as per Figure 13.

The hexaflash circuit can be rearranged to give uniform steps by getting rid of the awkward factor of three in the range. Although one option might be to count in units of thirds of fringes rather than whole fringes, that option also reduces acceptable slew rates for phase unwrapping. A fuller solution comes by noticing

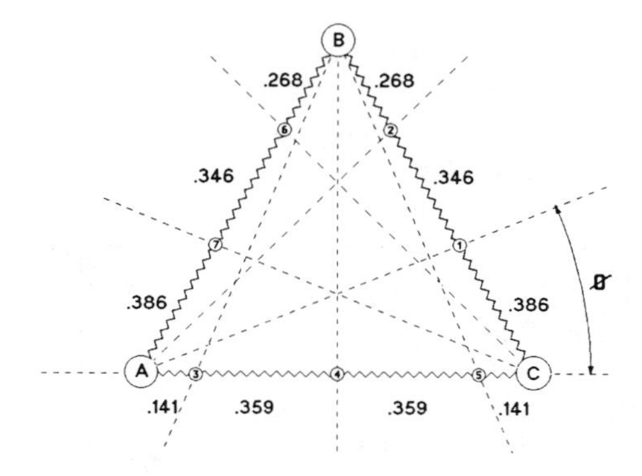


FIGURE 16. Polarflash network.

that if the $ABC$ signals are applied to a triangular network as shown in Figure 16, then a line through the triangle at angle $\phi$ intersects equal volatages. Taps can be organized along the network edges for each phase step, the taps being compared with the opposing vertex to ascertain at what angle the voltage balance occurs. Figure 17 shows a circuit of gates for 4-bit decoding of that balance. The ensemble of Figures 16 and 17 is named a *polarflash converter*. Low-resolution versions can

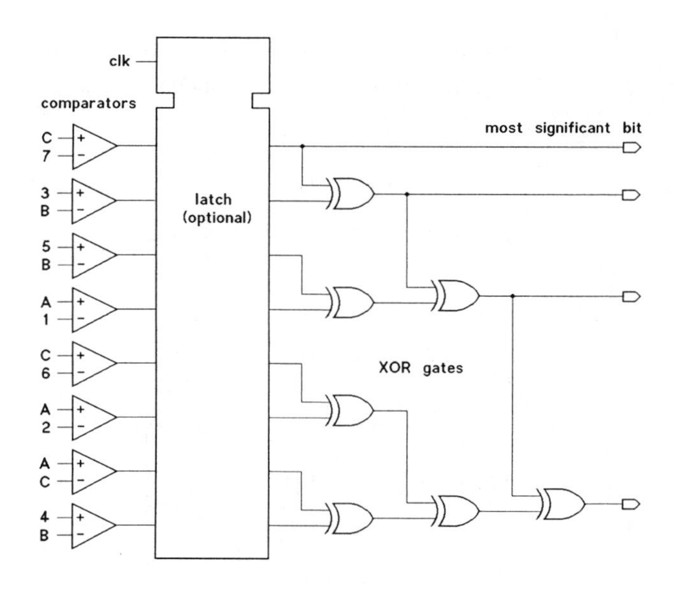


FIGURE 17. Polarflash decoding. Inputs refer to taps in Figure 16.

be built with commonplace chips, but custom, fully integrated circuit chips would be needed for 8-, 10-, or 12-bit resolution.

## Angle Encoder

Before pursuing further enhancement of phase metrology, I would like to describe an experimental demonstration of the capability so far. The demonstration employed an angle-sensing interferometer whose configuration is shown in Figure 18. It is basically an ordinary Michelson interferometer acting as the stator of the sensor, but whose beams have been folded a couple of times by the parallel mirror pair that act as the rotor. The remarkable feature is that the configuration remains in interferometric quality alignment for motions in all six degrees of freedom of the rotor. The path difference is independent of five of those degrees of freedom, but it is very sensitive to one of the angles of the rotor with respect to the stator. As a consequence, the performance is very robust. There is no need for a precision bearing or for any bearing at all; the device even operates while the rotor is being hand-held. That quality is loosely dependent on the parallelism of the rotor mirrors.

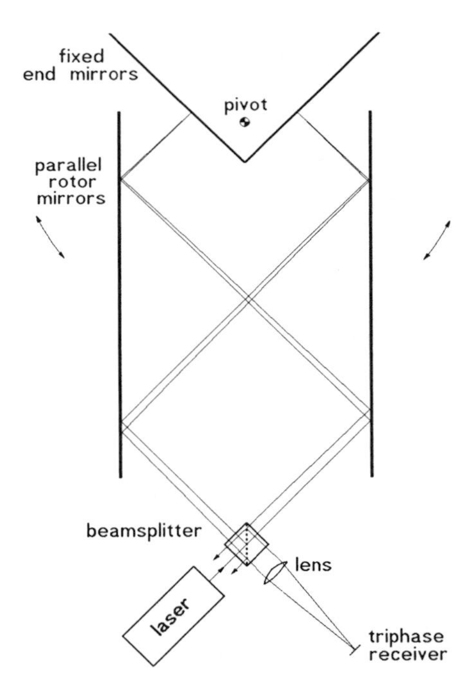


FIGURE 18. Interferometer for high-resolution tiltmeter angle encoding.

Empirically, they need not be parallel to Fabry-Perot tolerances, but nevertheless to better than one degree. For a spacing of 12 centimeters between the parallel mirrors and the HeNe wavelength (633 nanometer), there are over a million fringes per radian motion.

Notice in the figure that the alignment is such that the return beams do not become colinear, as is customary to get bull's eye fringes, but are separated by several millimeters. They are also arranged so that they straddle the input beam. There are a couple of big advantages to this arrangement. For one thing, they do not go back into the laser source, so no Faraday isolators are needed. Lasers get very upset when they receive external feedback from their own light. The arrangment would also make it possible to increase the optical efficiency by a factor of about four by making the beamsplitter a narrow semitransparent stripe between fully reflecting and fully transparent regions so that all of the return light would get directed toward the detector system. That not only doubles the light but also avoids the 50% absorption loss of the recombination. The lens finally brings the separated beams together at a small angle to provide the small patch of parallel fringes for the lenticular receiver.

Figure 19 shows the actual mechanical construction of the interferometer arranged vertically as a sensitive tiltmeter. The rotor, hung by a pair of strings, has

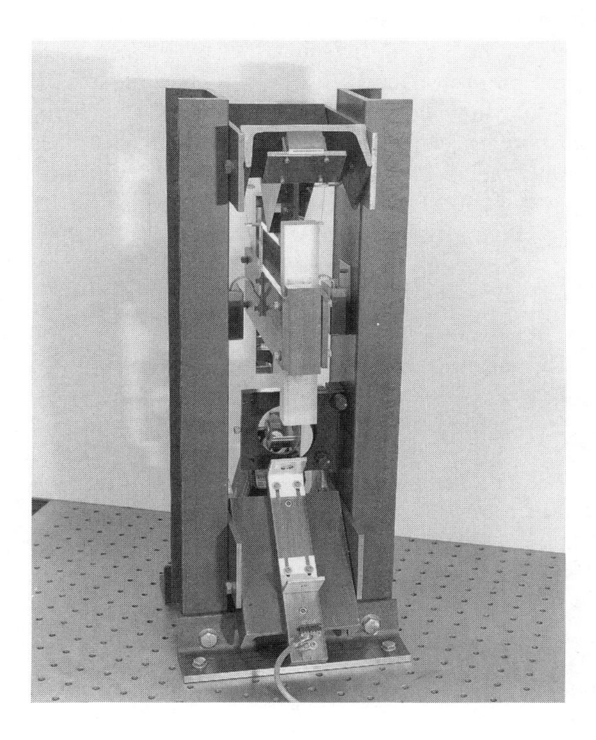

FIGURE 19. Tiltmeter angle-encoding interferometer.

a couple of paddles that dip in to pools of molasses for damping. The molasses has been thickened to the consistency of soft tar by boiling and is covered by a thin layer of oil. The step function response of the tiltmeter is shown in Figure 20. There are two principal components in that response: the exponential decay of the overdamped fundamental mode of the pendulum and the oscillations of the underdamped second (zig-zag) mode. Figure 20 was made possible through an interface acquiring 32 bits from the accumulator to a desktop computer. The upper 12 bits represent the integral fringe count and the lower 20 the fractional count. Figure 18 displays 64 fringes full-scale vertical from 10 intermediate bits subsampled at 0.01 second per sample, the filter clock rate being 2 MHz. There was no averaging involved ($N = 1$). The computer simply skipped over 20,000 primary samples between each accepted sample.

More magnified traces such as those of Figure 21 of phase versus time show better the capabilities of the system. The vertical scale is magnified to 1/64th fringe full scale by displaying lower order bits. At this scale the trace wraps. The horizontal scale is magnified by having the computer sample more quickly, in this case 5376 samples per second. Figure 21A shows no averaging ($N = 1$), so the computer is now skipping over 372 primary samples between each accepted sample. The hexaflash resolution is evident as the steps. Figure 21C shows a comparable trace with on-the-fly averaging of effectively 2048 samples ($N = 2048$), so the trace is much smoother with the penalty having a proportionately reduced maximum slew capability. Much of the smoothness is accountable to the 5.5 times oversampling even at the 5376 seconds$^{-1}$ rate. Figures 21B and D look at other data bits, D showing much more phase detail in the stable regions, but at the expense of much more frequent display wrapping. Inasmuch as there are more than a million fringes per radian, the sensitivity of the tiltmeter is clearly approaching the picoradian level, at 500 Hz bandwidth! A picoradian is a remarkably small angle, subtending only half the thickness of a credit card at the distance of the moon.

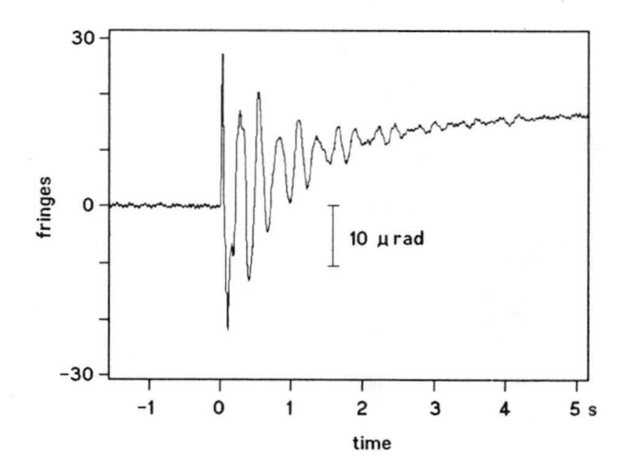


FIGURE 20. Dynamic step response of tiltmeter.

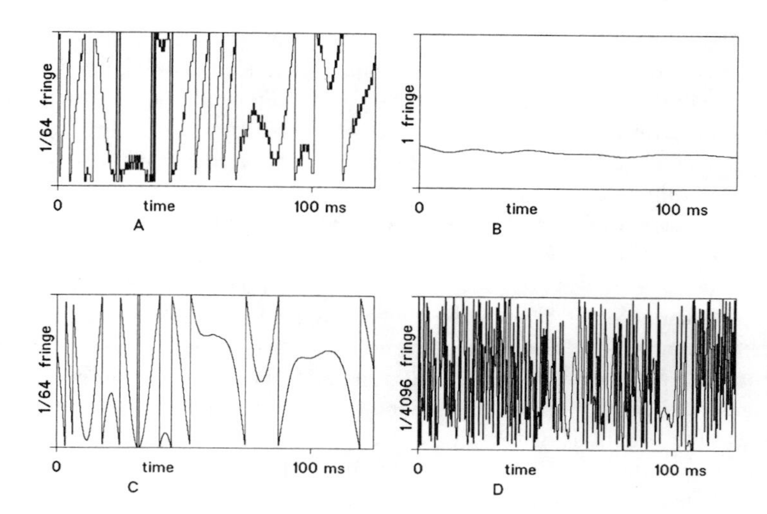


FIGURE 21. Quiescent performance of encoder. Note the absence of noise in D, where display is not wrapping and the signal is most steady. *A:* No averaging showing raw quantization steps. *B:* $N = 2048$ averaging, 921 nrad full scale. *C:* Same curve as B except 14.25 nrad full scale. *D:* Same curve as B except 223 prad full scale.

Such sensitivity is comfortably close to the shot noise expected for the half-milliwatt laser. The shot noise estimate is obtained as follows. Inefficiencies of the beamsplitter and detector reduce the effective laser power to about 0.1 milliwatt, which is about $3(10^{11})$ photons per millisecond. Each photon is measured to a third of a fringe. The rms uncertainty thus gets reduced to about $0.6(10^{-6})$ fringe when averaged over a millisecond.

The delivered sensitivity shown in Figure 21D is considerably better than had been expected. An understanding of how that might come about opens further opportunities. In the interpretation of measurements we ordinarily deal with both random and systematic errors. Random errors are expected to scale in proportion to the square root of the bandwidth, whilst systematic errors cannot get averaged away. This latter presumption turns out to be amiss, and its denial leads to the strategy of oversampling.

## Oversampling

Consider the error spectra. Random errors exhibit a white power spectrum. So if the bandwidth is limited, the noise power goes in proportion with the bandwidth and the root mean square noise goes as the square root of the bandwidth.

Systematic errors, such as those that are introduced upon quantization of a signal, are not supposed to change, and so their power resides at dc. Curtailing the bandwidth would have no effect. Suppose, however, that a sawtooth dither having a peak-to-peak of one quantization step were added to the signal. Curtailing the bandwidth by averaging $N$ samples during a dither period now reduces the quantization error by a factor of $N$, which is significantly better than $\sqrt{N}$. It is this fundamental ability to improve the scaling relation that makes oversampling strategies worthwhile.

Nevertheless, in this example the error spectrum is mainly at the dither frequency, which still may be rather low, encroaching toward the dc signal. That situation can be improved easily however by rearranging the $N$ samples. For example, we might negate alternate samples, making the sawtooth appear as an envelope on a Nyquist carrier frequency. Now the error spectrum lies up near the Nyquist frequency, much more separated from the dc. The separation is even more pronounced for a triangular (diamond), rather than sawtooth, envelope on the Nyquist frequency. As long as the amplitude distribution of the dither waveform uniformly fills a step of quantization, the averaging benefit remains unaltered.

Most of the recent popularity of oversampling, such as is found in digital audio, centers on delta-sigma data converters. They use closed-loop feedback to shape the dither waveform so that quantization errors are moved to high frequencies, where they can be removed thoroughly by low-pass filtering. It seems simpler to me, however, to generate the dither waveform independently and add it open loop to the signal.

A particular waveform that I find appealing may be called *bit-reversal*. It is generated by clocking a binary counter with the sample frequency and applying the output bits in reversed order to a D-A converter. It is the same type of bit-reversal that is employed in the Fast-Fourier-Transform algorithm. Although its spectrum is not so exclusively near the Nyquist frequency as that of the diamond envelope waveform, it seems to perform slightly better, as will be examined a little later. One period of both of these dither waveforms and their power spectra is shown in Figure 22.

The next stage is how to introduce the dither onto the three-phase signals. While it would be possible to physically dither the path difference of the interferometer, that would require high-frequency detectors with their concomitant wide noise bandwidth. It is much more preferable to have slow optical detectors and subsequently impose the dither electronically as a phase modulation. This turns out to be rather easy after noticing that the difference of any two of the $ABC$ phase signals is effectively a vector at right angles to the third signal and having a length proportional to the amplitude $\beta$. Three circuits, one for each phase, like the one shown in Figure 23 serve very well to apply any dither waveform as a pure phase modulation on our signals.

The next stage is to try to understand, develop, and design a good dither waveform so as to shape the error spectrum toward high frequencies and minimize any residual bleeding of that spectrum into the very low signal band. An experimental test interfaced to a computer has helped to gain some insight along these lines. The test signal is a very slow linear drift of the phase. Its spectrum is dominated by the first few harmonics of the entire test duration. The difference between the test

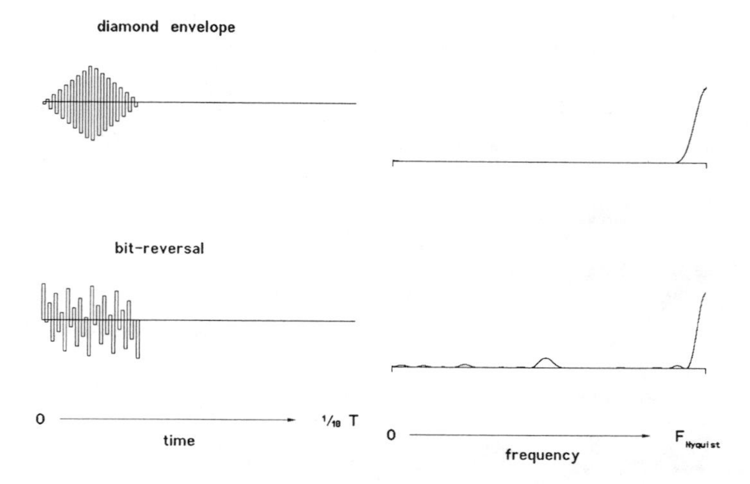


FIGURE 22. Two dither waveforms and their power spectra.

drift and its quantized representation (a staircase) is a sawtooth waveform whose spectrum still is composed mainly of low frequencies, as is shown at the top right of Figure 24.

Lower portions of Figure 24 show the consequences of having added two different dither patterns. Where the error is chattering at the Nyquist frequency, the upward and downward traces overlap in the figure to appear as solid black. For the diamond envelope dither the error appears as bursts of Nyquist frequency. While most of the spectrum is up near Nyquist frequency, there is a nasty amount down at low frequency from the burst envelope. For the bit-reversal dither, the error shades rather than bursts to Nyquist frequency. The result is a big lump in the spectrum at half-Nyquist frequency, but somewhat less contamination near zero frequency. In that fashion the error has been tailored for removal by low-pass digital filtering. There is a question as to whether residual leakage will be due to residual error at

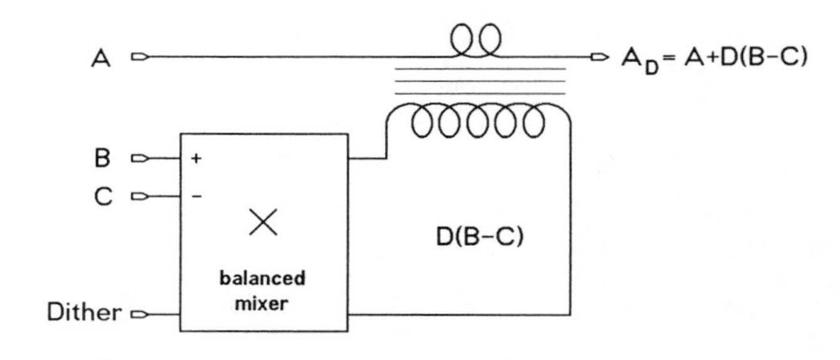


FIGURE 23. Phase modulator.

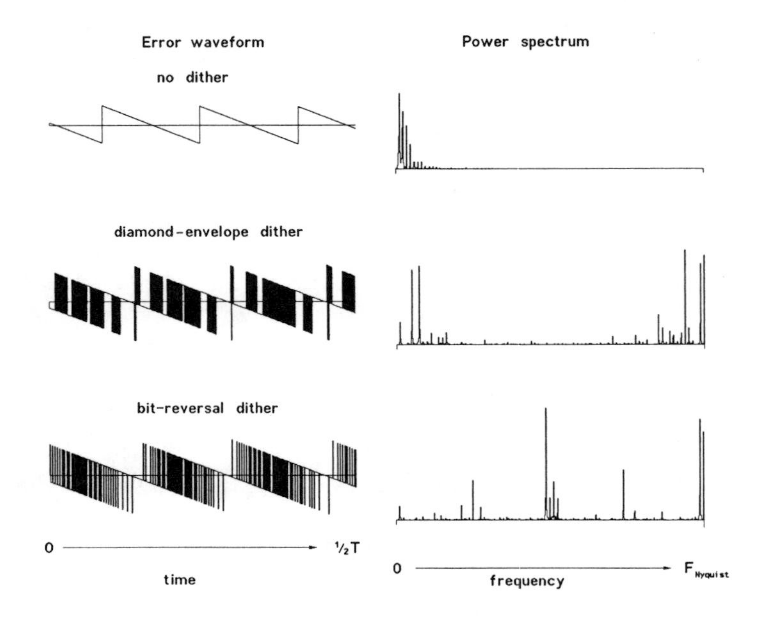


FIGURE 24. Quantization errors and their power spectra for dither schemes.

very low frequency within the passband of the filter or due to imperfect out-of-band rejection by the filter at half-Nyquist frequency. It looks to me as if the latter will dominate.

The previously described recursive phase unwrapping filter is first-order digital low-pass. Its rejection ratio at half-Nyquist frequency is linearly proportional to its decay constant $N$, which is also its bandwidth reduction factor. Hence we can expect the same noise reduction when it is used in conjunction with bit-reversal dithering.

Much better rejection is available from second-order filtering. Such filters have become quite feasible with DSP (Digital Signal Processing) chips based on VLSI technology. Their design and behavior are thoroughly treated in DSP literature, so only a brief introduction is due here.

## Second-oder Filtering

Second-order recursive filters have the form

$$\Phi_{t+1} = B_0\phi_{-t} + B_1\phi_{t-1} + B_2\phi_{t-2} - A_1\Phi_t - A_2\Phi_{t-1} \ .$$

(The first-order filter already treated is a reduced case where $B_0 = N^{-1}$, $A_1 = (1 - N^{-1})$, and $B_1, B_2, A_2 = 0$.) The second-order transfer function is expressed by

$$H(z) = \frac{B_0 + B_1 z^{-1} + B_2 z^{-2}}{1 + A_1 z^{-1} + A_2 z^{-2}} ,$$

where $z = \exp(2\pi i \ f/f_s)$, with $f$ being the frequency and $f_s$ the sample frequency. Literature on digital signal processing deals with how to determine the $B$, $A$ coefficients. It should be clear that for unity dc gain $(H(1) = 1)$, $1{+}A_1{+}A_2 = B_0{+}B_1{+}B_2$. For Butterworth filters, $B_0 = B_2 = \frac{1}{2} B_1$.

Straightforward direct calculation of the recursion relation often leads to dynamic range problems that can be alleviated by adding and subtracting $A_2 \Phi_t$ and then grouping the terms to give

$$\begin{aligned} \Phi_{t+1} &= B_0\phi_t + B_1\phi_{t-1} + B_2\phi_{t-2} - (A_1{+}A_2)\Phi_t + A_2(\Phi_t - \Phi_{t-1}) \\ &= \Phi_t + (B_0\phi_t + B_1\phi_{t-1} + B_2\phi_{t-2} - (B_0{+}B_1{+}B_2)\Phi_t) + A_2(\Phi_t - \Phi_{t-1}). \end{aligned}$$

The first terms on the right resemble a simple first-order filter using a weighted sum of the three most recent inputs. The final term gives a dampening proportional to the difference of the two most recent outputs.

The power rejection at half-Nyquist frequency for these filters is evaluated as $H(-i)H^*(-i)$ since $-i = \exp(-\pi i/2)$ and where $H^*$ is the complex conjugate of $H$. For first-order filters such evaluations give a power rejection that is proportional to the square of the bandwidth reduction $N$. For second-order filters the power rejection goes as the fourth power of the bandwidth reduction. Hence the second-order filters should be much more effective in removing the residual quantization error occurring at half-Nyquist frequency for the bit-reversal dither situation.

This vast improvement of the scaling relation means that quantization error can be reduced below shot noise error even for lasers of significant power. A half-watt laser puts out about $10^{18}$ photons per second; accordingly, its shot noise should allow measurement to about a nanoradian per $\sqrt{\text{Hz}}$. Sampling to a millifringe at 2 MHz scales down to a nanofringe at 1 KHz bandwidth, distinctly below the shot noise level. Hence shot noise rather than quantization error will limit the measurement.

## Analog Comparison

How does this digital approach relate to the analog methods that generally are used for interferometric metrology? A popular technique is to modulate the phase, like a sinusoidal dither, and then look at the signal frequencies that emerge. The results are used to control a servoactuator that acts to shift the operating point so as to nullify the ac signal. Two alternatives may be used: one is small-amplitude modulation around the black fringe, with the rationale of keeping the intensity and its concomitant shot noise low, and the other is to look at the difference of complementary outputs from a beamsplitter, which in effect perceive the phase $\pm 90°$. The former alternative begets the problems of the operating point around the black fringe exhibiting minimum phase sensitivity, and that the fringe contrast must be excellent for the fringe to be really black. The latter alternative, balanced homodyne reception, is somewhat like the triphase system, except that it uses 90° rather than 120° shifts and has only two ports, so that it only works at the operating point where the signals are in balance. At other phases, the results are susceptible to intensity fluctuations.

The most demanding and timely application of oversampling metrology would be the Laser Interferometer Gravitational-Wave Observatory (LIGO). They are attempting to reach the shot noise limit for a 60-W laser (2 x $10^{20}$ photons/s) over a KHz bandwidth. That is a lot of power to handle, and leads to about 7 x $10^{-11}$ rad/$\sqrt{}$Hz rms or $10^{-11}$ fringe/$\sqrt{}$Hz. It should be possible to do flash conversion to a millifringe at 10 MHz, and reducing the bandwidth by $10^4$ in conjunction with dither and a second-order filter should bring the quantization error to $10^{-11}$ fringe over the KHz or 3 x $10^{-13}$ fringe/$\sqrt{}$Hz. Hence shot noise will dominate. However, it seems doubtful that sensitive detectors can dissipate such power, and thus it would probably be a good idea to reduce the power and bandwidth.

A separate issue would be to recommend a 60° rather than a 90° included angle for the interferometer beams. 60° permits three interferometers around an equilateral triangle of beams, so that the observatory would have a broader response pattern and be able to sense direction as well as the existence of gravitational events. Furthermore, the three interferometers would offer extra redundancy for closure information to help exclude seismic contamination. Haste may have already precluded such an option, however.

Numerous advantages would accrue to the digital oversampling approach as compared with analog servomeasurement:

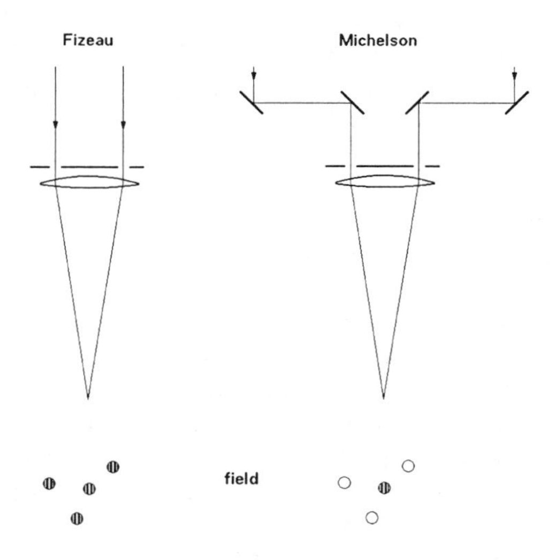


FIGURE 25. Comparison of aplanatic Fizeau stellar interferometer with nonaplanatic Michelson stellar interferometer.

1. Mechanical and optical simplicity.
2. High sensitivity to reach shot noise, just as for analog.
3. Good quantum efficiency with all of the photons participating in each measurement. Admittedly, however, the modulation efficiency drops to 83% if the detector has no gaps at all.
4. Subkilohertz performance, where the prospects are better for finding gravitational waves.
5. Insensitivity to laser fluctuations. Laser intensity noise forces systems having fewer than three simultaneous signal ports to operate at excessively high frequencies.
6. Direct time-domain operation retaining the phase information that distinguishes events as transient. Power spectra may suffice as diagnostics, but not as measurements.
7. Works at any operating point of the fringes.
8. No hysteresis.
9. Self-calibrating.
10. Totally passive with no piezos or servos to complicate the dynamic behavior.
11. Digital with unlimited dynamic range.
12. Permits numerical exclusion of seismic interference. Since seismic excitation acts on the suspension, whereas gravitational excitation acts directly on the test mass, a compound pendulum can be used to obtain two measurement channels. A linear combination of the data from these channels can be used to exclude the seismic contamination in real time.
13. Robust operation, wherein the behavior only degrades slowly for nonideal circumstances.
14. High-bandwidth recording for a posteriori analysis.
15. Better opportunities for commercial spinoff to metrology interferometers, lidar, fiberoptic interferometric sensors, optical frequency standards, and gyroscopes.
16. More economical. The costly digital technology development already has been done, paid for, and amortized.

It is for analogous reasons that digital compact discs have replaced analog long-playing records in high-fidelity audio.

## Stellar Interferometry

My original motivation for this effort had been that getting the phase might lead stellar interferometry toward optical aperture synthesis. As it turns out, that task entails much more awkward geometric, chromatic, and atmospheric problems than does laser interferometric metrology. For example, the geometry of the interferometer should be such that its apertures behave as if they were subapertures of an immense high-quality telescope. That means not only the matching of time delays between the two apertures, but also within those apertures if we hope to achieve

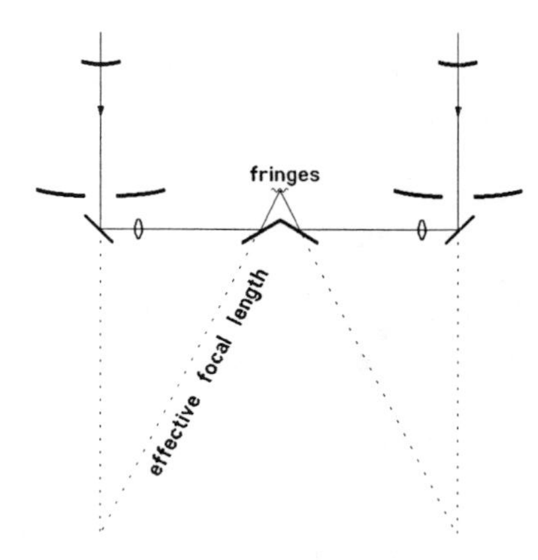


FIGURE 26. Beam-combining of Cassegrain apertures for aplanatic interferometry.

interference over some field of view. Such is the circumstance of the Fizeau stellar interferometer, but it is not the case for the Michelson stellar interferometer, as shown in Figure 25. This situation was recognized as pupil matching early in the conception of the Multiple-Mirror Telescope, and it is identically the same as aplanatism in optical design, as was encountered back in Chapter 1. Figure 26 shows how aplanatism may be accomplished in the combination of two separate Cassegrain apertures.

Chromatism poses another problem. The most interesting sources are already too faint to afford narrow-band filtering. The fringes might be achromatized with a dispersive prism such as the one shown in Figure 27, but the solution clearly upsets the aplanatic conditions. It is probably more satisfactory to employ crossed dispersion so as to obtain a two-dimensional display of fringe delay along one axis versus wavelength along the other, and then sort them out using a fully two-dimensional CCD sensor. All of these complications along with atmospheric turbulence and earth rotation thus have discouraged my hoped-for astronomical use of the triphase technology.

## More Applications

On the other hand, the spinoff to more mundane applications is looming. Polarization interferometers can be readily adapted to the three-port mixer simply by splitting the output with a piece of calcite, as shown in Figure 28. It is essential to get the beams separated, as with a shearing interferometer, and then to bring them together at angle with a lens. Recalling Figures 5 and 14, the two beams can even come from different lasers for accurately comparing their phase differences

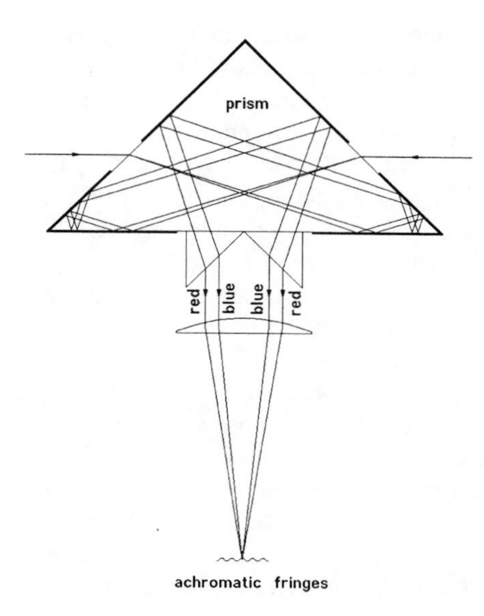


FIGURE 27. Achromatic beam-combining that unfortunately renders the operation nonaplanatic.

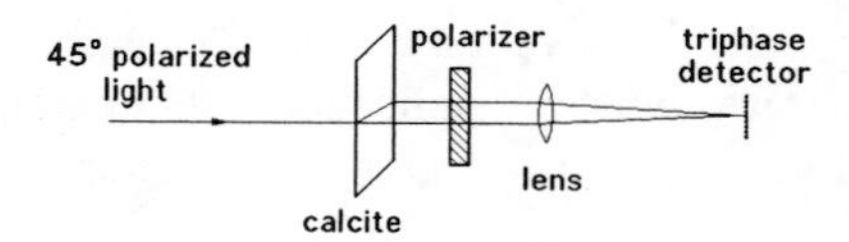


FIGURE 28. Polarization mixing to three ports.

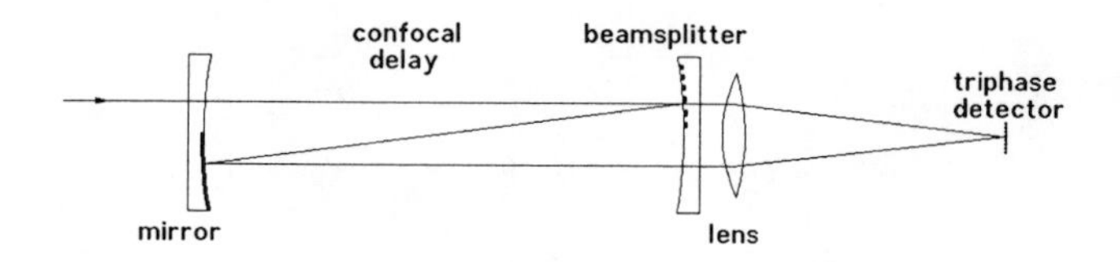


FIGURE 29. Delayed self-homodyne interferometer.

or frequencies. The comparison also might be with a passive cavity of confocal form as shown in Figure 29 for delayed self-homodyne operation.

Frequency drifts of the laser lead to phase shifts of the fringes, and because we can measure very quickly to a small fraction of a fringe, it should be possible to monitor the laser frequency to $\Delta f \, \Delta t \ll 1$. With that capability I am led to wonder just what happens during a longitudinal mode-hop of a laser. With HeNe lasers in a longitudinal magnetic field, for example, the Zeeman beat frequency peaks when the circularly polarized modes just straddle the Ne resonance frequency. As the laser warms up, its cavity length increases and the Zeeman beat frequency undergoes periodic peaks as shown in Figure 30. These are quite reproducible, and the nature of the glitches that occur on the shoulders of the peaks remains a mystery to me. It would be helpful to try following the actual laser frequency rather than the Zeeman beat through those variations. Frequency monitoring for semiconductor laser diodes should be just as interesting.

This very same arrangement is eminently suited for Quantum Non-Demolition (QND) measurements. One technique for QND involves two or three optical solitons. Two of them serve as a pair, with one as a reference either leading or lagging the sensing soliton by a fixed time delay. The third, or signal, soliton is arranged to collide at an angle with the sensing soliton, in which case it shifts the phase of the sensing soliton with respect to the reference soliton. There is no demolition of photons in the collision. Hence the phase shift serves to sense the presence or absence of the signal soliton in a QND fashion. By making the cavity delay of Figure 29 match the delay between the signal and reference solitons, they will arrive together to make fringes for the phase sensing mixer.

There are also noninterferometric applications; for example, we may consider a noncoherent optical radar. A three-stage ring oscillator provides a $3\phi$ local os-

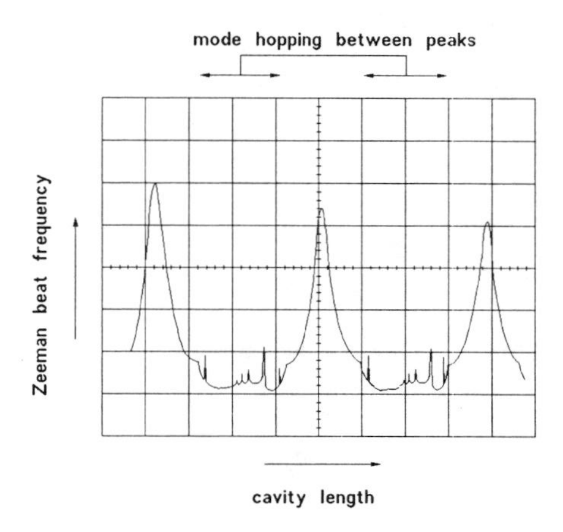


FIGURE 30. Mode-hopping activity as a Zeeman HeNe laser warms up. The glitches are reproducible but unexplained.

cillator. One stage is used to modulate the transmitter laser. The return photodiode signal then may be mixed with each of the three phases for the hexaflash circuit.

One might also consider an optical "selsyn." Imagine three stationary slots radial to a rotor axis. An eccentric circular disc mounted on the rotor serves as a shutter for the three slots at 120°. Illuminating the slots diffusely with LEDs and detecting the light with photodiodes provides the three phase signals for subsequent processing. The result is like that for a selsyn, except that there is no need for ac excitation.

## Quantum Ramifications

All of this measurement technology supports the original assertion that phase and position are one and the same variable. In quantum mechanics the Fourier transform conjugate variable of position is momentum or direction. That relation is quite clear from the Young's fringe experiment in Figure 31. When the fringes are resolved it is impossible to know through which slit the photon passed, because the momentum is too uncertain to tell whether the photon is going upwards or downwards. The Fourier transform conjugacy of the variables is even more explicit when a lens is introduced as in Figure 31. The assertion is reinforced by the common occurrence of the phase variable $(\omega t - kx)$ in the solution of the wave equation. For $x = 0$ we are left with a function of $t$, but for $t = 0$ we are left with a function of $x$. Measurement processes virtually always mix two beams to make standing waves that leave phase primarily as a function of $x$.

But if phase and momentum are conjugates subject to the Heisenberg Uncertainty Principle, then there must be something wrong with the assertion that phase and number states are conjugates. This latter assertion commenced with Dirac's specification of a phase operator. Minor inconsistencies were revealed in that specification, but the equivalent cartesian expressions of sine-cosine, real-imaginary, or creation-annihilation operators were deemed valid. That story is told by Carruthers and Nieto in their review article "Phase and angle variables in quantum mechanics"

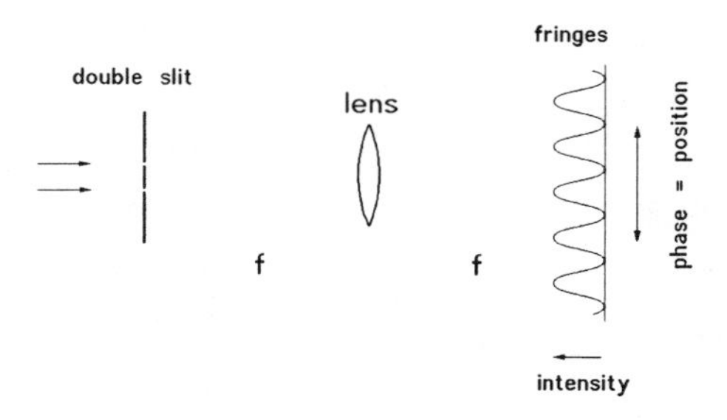


FIGURE 31. Young's experiment with Fourier transform lens to emphasize conjugacy of slit and fringe planes.

[1968], wherein the acceptance of the cartesian expressions derives from associating them with a harmonic oscillator. I will now argue that that association is unfounded.

For unidirectional wave propagation there are no negative frequencies. Hence the antisymmetric part of the spectrum must exactly cancel the symmetric part for negative frequencies. That implies that the real and imaginary components of the wave are Hilbert transform conjugates. The situation is complementary to that of the Kramers-Kronig dispersion relations that result from causality. In that case, the response must be zero for negative times, so the real and imaginary parts of the refractive index must be Hilbert transform conjugates. While Hilbert transform conjugates do bear some resemblance to a derivative, they are certainly not Fourier transform conjugates. Furthermore, when there is a bidirectional superposition of waves, as is the case for mixing, then even the Hilbert relationship does not hold.

On the other hand, the accepted conjugacy of the creation-annihiliation operators has served as the basis for numerous reproducible experiments indicating squeezed states of light. The tail end of those experiments usually has a piezo phase actuator, a balanced homodyne mixer, and an rf spectrum analyzer as shown in Figure 32. This is essentially the same as the output end of a Mach-Zehnder interferometer. What the experiments show is that for squeezed light the noise level, as measured by the spectrum analyzer, dips below the shot noise level twice per fringe as scanned by the piezo. The spectrum analyzer is running as a high-pass filter, and it would have been extra helpful to see its output versus its input. If the result were concave upward, then the dips of noise would have been at nodes of the fringes; whereas if it were convex downward, the dips would have been at the peaks of the fringes. Recall the same distinction made earlier in this chapter concerning the noise versus fringe phase effect in balanced homodyne reception for intensity fluctuations as compared with flip-flop ratiometry. Unfortunately, the output for the squeezing experiments was shown only as a function of the piezo voltage, so the phase of the noise dips relative to the fringes is unclear. Nevertheless, the experiments were quite reproducible, verifying the relevance of the Uncertainty Principle. I venture to suggest that it may not have been the Heisenberg Uncertainty Principle but merely a statistical uncertainty relation that is less compelling than Heisenberg's.

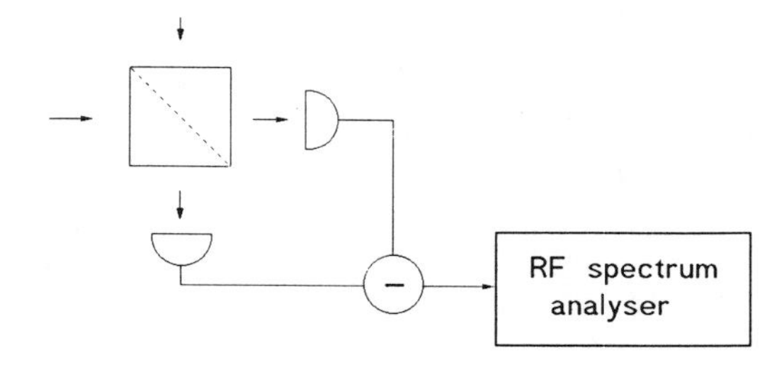


FIGURE 32. Spectrum analyzer on balanced homodyne receiver, as in squeezed light experiment.

Suppose that we construct a vector composed of $N \pm \sqrt{N}$ unit photon vectors at random phase angles. The resultant typically will be $\sqrt{N}$ long. But if there is a systematic phase, according to the contrast $\Gamma$ of fringes, the resultant will decompose to a vector $N\Gamma$ long having the systematic phase angle plus a random vector $\sqrt{N}$ long as shown in Figure 33. Although some factors of two have been omitted for simplicity, this diagram leads to a statistical uncertainty between the length and phase angle of

$$\Delta n \, \Delta\phi \;=\; 1/\Gamma \;\geq\; 1 \,,$$

since the contrast $\Gamma$ is necessarily between zero and unity. There we have a statistical uncertainty that resembles Heisenberg's. If our fringe observations are projections of that Argand diagram, as is often presented, then we need not claim Heisenberg's relation. It is not as simple as that, however. Projections of Figure 33 cannot represent observations because the photons then would contribute fractionally rather than as integer quanta. The true description remains to be settled.

Another approach to fringe sensing at the quantum level is from photon correlations. In the early 1960s the interpretation of Brown-Twiss bunching of photons was a controversial topic that became even more controversial when Glauber argued that laser light should not exhibit the bunching. Gamo explained it to me as follows. For ordinary narrow-band light, that bandwidth results from both amplitude and phase modulation side bands. When we detect a photon we are probably near an amplitude crest, and so there is an excess probability for detecting a second photon. The photons are thereby bunched. Actually, the band must be quite narrow for the crest to last long enough for us to notice the bunching. Lasers, on the other hand, behave as oscillators, so their bandwidth comes exclusively from phase fluctuations. There are no crests of amplitude and so there is no bunching, and Poisson statistics apply. A seemingly trivial complementary situation is given by Cerenkov radiation or synchrotron radiation. There we have shocks or pulses of electric field amplitude. If these are filtered through a Fabry-Perot etalon, the etalon rings, but

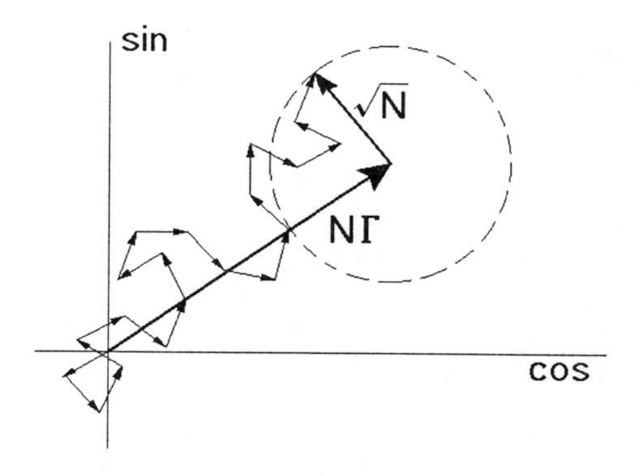


FIGURE 33. Heuristic Argand diagram for a fringe composed of $N \pm \sqrt{N}$ photons.

the bandwidth comes from primarily amplitude fluctuations. Hence the way we sense those radiations is as pulses, or very bunched light.

Now consider mixing two laser beams to give fringes at the beat or difference frequency. These fringes are periodic amplitude fluctuations and so can be sensed by photon correlations, just like Brown-Twiss bunching. For example, two detectors can be separated by one fringe in a pattern, so that if we sense a photon in one detector, it is probably at a crest and the other detector is at the next crest, and so photon correlations occur. On the other hand, if the spacing is half a fringe, there will be anticorrelations. There need be no special quantum mystique about those phenomena of photon correlations that result merely from synchronously moving fringe patterns.

Another statistical case is that of regularly pulsed light, such as that generated by a mode-locked laser. Sensing a photon ensures that we are at a pulse, and we thereafter can be sure that there will be no photons between subsequent pulses. That much can be said with confidence about the photon statistics, but suppose that the beam is faint enough so that there is only about one photon per pulse. Will there then be a regular sequence of photons? In most instances probably not, because the pulses themselves will have a Poisson distribution in the number of photons per pulse. On the other hand, there are experiments with specific sets of atoms and with Optical Parametric Oscillators that lead to predictable detections of photons, at least when the detectors have close to perfect quantum efficiency. Such departures from Poisson statistics may be expected as a result of the conflict of detailed conservation of energy with the very assumption of Poisson statistics that the detection of each photon exercises no influence on the detection of other photons. For a limited total number of photons, the detection of each photon ought to diminish the probability of detection of subsequent photons. But the formation of interference fringes at light levels so low that there is virtually never more than one photon at a time in the apparatus implies that the photons are not localized in flight. Consequently, the detection of a photon over here cancels the probability of detection over there, action at a distance that may occur faster than the velocity of light. I certainly am unable to venture a resolution of the dilemma.

## Bibliography

A. Abramovici *et al.*, 1992 "LIGO: The Laser Interferometer Gravitational-Wave Observatory" *Science* 256: 325–333

T. Baer, F. V. Kowalski, and J. L. Hall, 1980 "Frequency stabilization of a 0.633 $\mu$m He-Ne longitudinal Zeeman laser" *Appl. Opt.* 19: 3173–3177

P. Carruthers and M. M. Nieto, 1968 "Phase and angle variables in quantum mechanics" *Rev. Mod. Phys.* 40: 411–440

K. Creath, 1988 "Phase-measurement interferometric techniques" in *Progress in Optics, V. 26*, ed. E. Wolf, Elsevier, pp. 349–393

P. Hariharan, 1985 *Optical Interferometry,* Academic Press

———1992 *Basics of Interferometry,* Academic Press

M. Hercher, 1991 “Ultra-high resolution interferometric sensors” *Optics & Photonics* (November): 24–29

L. Mertz, 1983 “Complex Interferometry” *Appl. Opt.* 22: 1530–1534

———1984 “Phase estimation with few photons” *Appl. Opt.* 23: 1638–1641

———1988 “Complex homodyne reception from discrete photons” *Appl. Opt.* 27: 3429–3432

———1989 “Optical homodyne phase metrology” *Appl. Opt.* 28: 1011–1014

———1991 “Interferometric angle encoder” *Rev. Sci. Instr.* 62: 1356–1360

Motorola *literature*, 1991 “Implementing IIR/FIR filters” APR7/D

W. H. Steel, 1967 *Interferometry,* Cambridge U. Press

D. F. Walls, 1983 “Squeezed states of light” *Nature* 306: 141–146

# 4

# Image Sensing

For many years photographic film served as the mainstay of observational astronomy. It is presently being supplemented by the Charge-Coupled Detector (CCD), especially at the limiting magnitudes. The main reasons are much better quantum efficiency and much better linearity, which lead to more sensitive, more quantitative observations. An additional reason is that the data are directly accessible to computers for all sorts of image enhancement and massaging. Economy and freedom from the nuisance of wet processing also play as factors.

## Imaging Photon Counters

A less known electronic detector is the imaging photon counter, of which there are various forms. The first to be employed used an intensifier tube in front of a television camera, and then an image processing algorithm to identify photon-induced events in the television frame. Although this strategy does work, I prefer to exclude it here because it fails to tally the temporal sequence of the photons found within the frame. Literal imaging photon counters record the digital *xy* coordinates of each photon as it arrives and so generate a sequential list of those coordinates. If this list is played backed through digital-to-analog converters onto an *xy* oscilloscope, the observation is reenacted photon by photon. Straightforward image accumulation can be done by incrementing pixels according to the list, but it is also possible to compensate against image drifting by shifting the coordinates just prior to pixel incrementing. In other words, it is easy to guide the telescope a posteriori to produce the sharpest image.

The front end of these imaging photon counters always includes a photocathode followed by several microchannel-plate electron multipliers in a vacuum envelope. A photoelectron from the cathode goes to the multiplier plates and results in about one to ten million electrons emerging from the final plate to a special anode that permits position determination. A variety of anode forms are feasible. The simplest is a carefully tailored resistive sheet that has four corner electrodes. The ratios of the charge pulses at these electrodes then serve to determine the centroid coordinates for the electrons. The main problem for this RANICON detector tube is that the anode resistance combines with inevitable capacitance to slow the response speed and limit the maximum count rate. Another form of anode, called the *wedge-and-strip,* is composed of three intricately patterned conductors that can be used to alleviate the resistance problem, although it requires accurate geometry. A third alternative is many anodes, although that entails many preamplifiers. That many-anode solution is called the MAMA detector. Yet another variation uses a phosphor anode with optical position sensing. The phosphor anode tube is a commercial product for military night-vision goggles, although a special fast phosphor is desired for the photon counting. Optical position sensing is done by making multiple (19) images of the phosphor anode onto grid masks with photomultipler tubes. The masks yield a Gray code of the spot position, and the whole system has been named, with a sense of humor, the PAPA detector.

All of these imaging photon counters work quite well. They have two important advantages as compared with CCD detectors. First, they keep a dynamic record of the observation and thereby allow dynamic image processing. Second, they exhibit much lower dark noise. A good CCD will have about 10 electrons of uncertainty per pixel per readout. That corresponds to a dark level of 100 electrons and discourages frequent readouts. In comparison, the imaging photon counters generally have less than one dark count per pixel per hour. Thus, even for significantly lower quantum efficiency, at very low illumination the photon counter can recognize fainter sources. The main problem with the photon counters is that they are far too expensive for most uses, and so the market is too small to lower the prices.

What is really needed to get by that impasse is an all solid-state imaging photon counter. To that end I had proposed the following configuration. Imagine a cross-point array of photothyristors. A *thyristor* is a bistable four-layer diode. Normally it is nonconducting, but it can be triggered to a conducting state. The central two layers can be fabricated as an avalanche photodiode for the triggering. If the array is mildly cooled, there will be no free carriers to initiate conduction, even though the voltage exceeds the avalanche level. The array remains poised waiting for a photon to create an electron-hole pair to start the avalanche. That is just how an avalanche photodiode is used in the photon-counting mode. The idea was then to read out which row was connected to which column with a peripheral array of ordinary encoding diodes as shown in Figure 1. Then the voltage could be removed to reset the chip to await the next photon. It all seemed plausible, but there proves to be a fatal pitfall. Once a thyristor starts to conduct, it lights up, emitting photons that will trigger all of the other thyristors in the array. It really is not possible to shield them from one another, and the process is too fast to ever know which triggered first. For that reason it is not possible to operate arrays of avalanche photodiodes in the photon-counting Geiger mode. While it still might be feasible to concoct CMOS equivalent thyristors as shown in Figure 2 for the array because their conduction would not involve light-emitting recombinations, the problem would be to get the capacitance low enough for one electron to bring the voltage over the triggering threshold. At any rate, there has not been sufficient incentive to proceed any further.

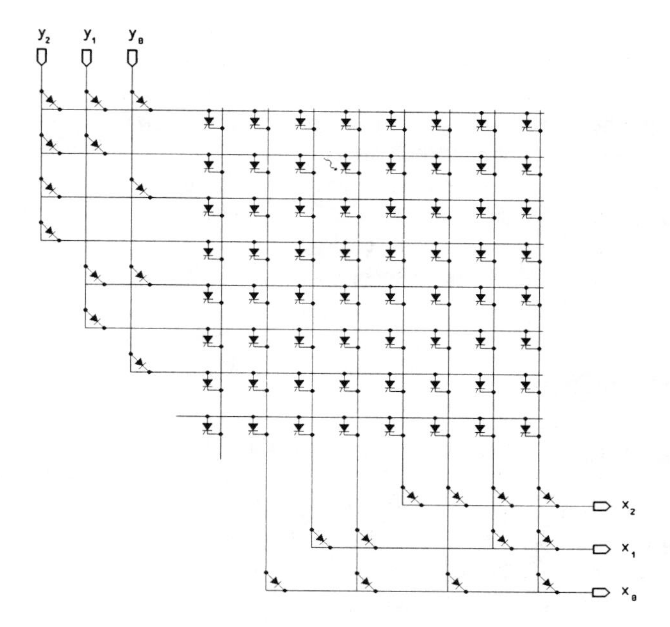


FIGURE 1. Photothyristor array.

# Heterodyne Detectors

Before leaving the topic of solid-state detectors, however, there is one other that might be mentioned. This has to do with heterodyne detection. Mixing light with a local oscillator gives a beat frequency that can be detected as an ac signal. This works fine for a diffraction-limited light beam, but not for a diffuse beam. The reason is that the mixture is speckled, with the speckles changing at the beat frequency rate. If the detector subtends more than one speckle, then the speckle

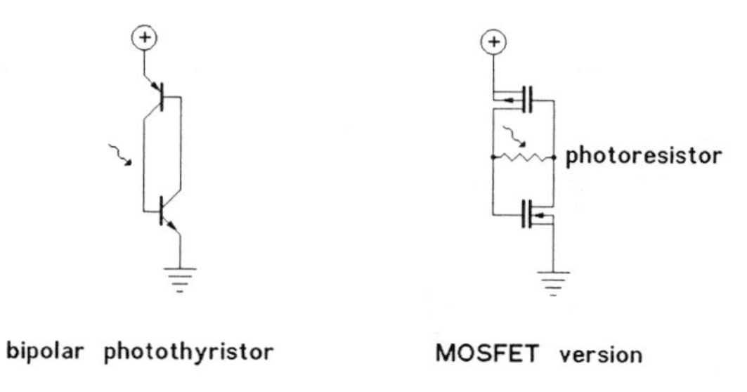


FIGURE 2. Bipolar and MOSFET equivalent photothyristors.

fluctuations tend to cancel out one another, leaving negligible overall fluctuation. But the small detector unfortunately wastes all that other light. A solution is to employ multiple detectors and sum their rectified signals. A convenient way to do this might be to make an application-specific integrated circuit composed of a two-dimensional array of cells as shown in Figure 3. The two photodiodes occupy separate pixels, and they pump charge onto the signal electrode when they have fluctuating signals. For an $N \times N$ array of cells the signal-to-noise should improve by $N$ as compared to a single cell. Using a tunable laser for the local oscillator would provide very narrow-band spectroscopy of diffuse light sources.

## Fourier Tracking

Now we can return to explore astronomical imaging with photon counters. Atmospheric turbulence sets the chief limit to the performance of optical telescopes by blurring the images. If it were just a question of the image jiggling around, we could cure that with a fast autoguider, but various portions of the telescope aperture jiggle differently from one another. In 1970, Labeyrie [1970] published a remarkably insightful article pointing out that short- exposure stellar images were dispersed into speckles, and that the motions of those speckles smoothly blurred long-exposure images. His idea then was to average cinematically the diffraction patterns of many short-exposure images to learn the average spatial power spectrum of the underlying star image. An unresolved star gives a broad spectrum, while a resolved star gives a narrower spectrum and a double star shows bands in the spectrum. Thus was born astronomical speckle interferometry.

The interferometry gets to the autocorrelation of star image and so it works fine for symmetric sources. Images ordinarily also include antisymmetric parts, and so it would be nice to deal with sine as well as cosine components. Notice that we have been led to a Fourier interpretation of images, just the same as with the Abbé theory of the microscope. If we consider the autocorrelation of the telescope aperture as in Figure 4, the shaded region contributes light to a specific Fourier component whose fringe spacing is inversely proportional to the aperture displacement shown in the figure. Atmospheric effects mainly shift the phases of the Fourier

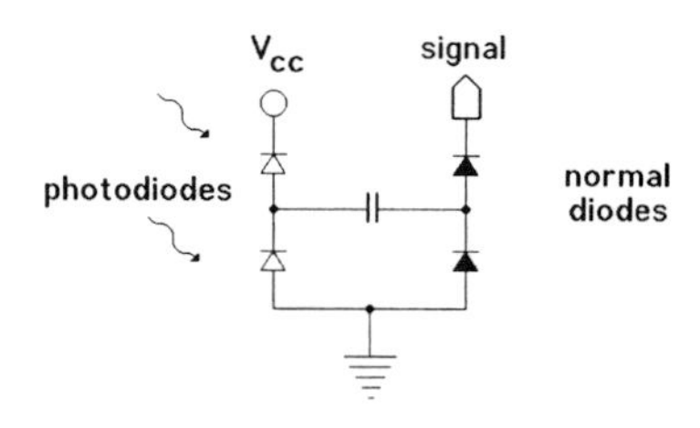


FIGURE 3. Heterodyne detection cell.

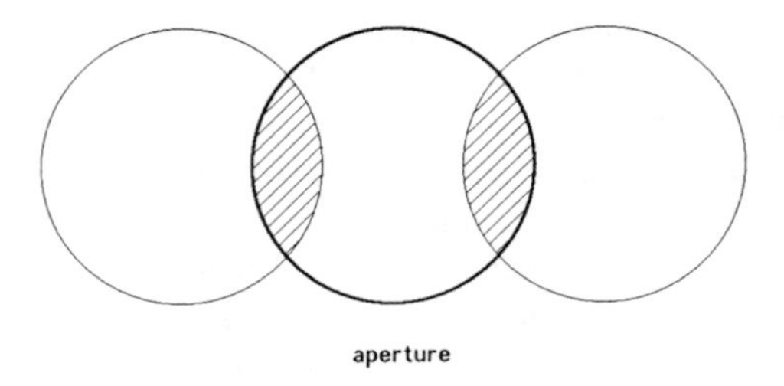


FIGURE 4. Autocorrelation of telescope aperture.

components, although when the shaded region is large there also can be a tendency for portions within the shaded region to cancel one another. For the latter reason there is sometimes a preference for a nonredundant aperture, whereby the shaded region remains small at all displacements and at the expense of total aperture area.

The atmospheric correction problem is mainly to correct the fringe phases and secondarily to restore their amplitudes. One idea, originally expressed by McGlamery [1970], has been to track the fringes so as to find an average phase and amplitude for each spatial frequency component. This contrasts with an ordinary long exposure where there is no tracking, so the fine fringes get drastically blurred and lose their amplitude. The problems with tracking are that the phase must be unwrapped continually and that accidental slipping of a fringe is catastrophic. These problems are especially serious at low signal-to-noise ratios, and that turned out to be a pitfall for the prevailing tracking methods.

The polar tracking filter described in the last chapter, on the other hand, shows much more tenacity than prevailing methods. The question is, does it have enough tenacity for the atmospheric correction problem? The analysis of some past image photon-counting records of stars suggests a tentative yes answer. This analysis simply tracks the effects of each successive photon upon each and every spatial frequency component of the image. The Fourier transform of an image composed by a bunch of photons is $\sum \exp i(ux+vy)$, where $x$, $y$ are the coordinates of the photons and $u$, $v$ are the spatial frequency variables of the transform. Thus, for each $x$, $y$ incidence of a photon we establish a phase $\phi = ux+vy$ as the input for a separate tracking filter for every $u$, $v$. In addition to the phase tracking filters, we maintain complementary amplitude filters:

$$\begin{aligned}
\Phi_t(u, v) &= \Phi_{t-1}(u, v) - (ux_t + vy_t) , \\
\Phi_{t+1}(u, v) &= \Phi_t(u, v) - N^{-1}\Phi_t(u, v) , \\
\Gamma_{t+1}(u, v) &= \Gamma_t(u, v) - N^{-1}\big(\Gamma_t(u, v) - \cos(\Phi_t(u, v))\big) ,
\end{aligned}$$

where $t$ is the temporal index of each photon and $N$ is an exponential decay constant number of photons. Since the phases $\Phi$ are unwrapped, we then can take long-term temporal averages of the phases $\Phi$ and the amplitudes $\Gamma$, convert these averages to cartesian formats (real and imaginary parts), and perform an inverse Fourier transform to reconstruct the tracked averaged image. The optimum $N$ is chosen by trial and error to maximize the values of $\Gamma$. In practice, $N$ will correspond to about 10-milliseconds worth of photons. Success will hinge on the absence of phase unwrapping errors.

Really suitable data have not been available for testing this procedure. The problems were that, although the imaging photon counter was obtained for this speckle imaging problem, it became available before this direct and tenacious phase tracking algorithm was devised so inadequate image scaling was used for the observations. Furthermore, the available 30-centimeter telescope was too small to display speckles even on bright stars. The relevant part of the testing was with respect to the low photon rate of 14,400 per second, limited to that maximum by the recording scheme.

Figure 5 shows the overall results of tracking $32 \times 16$ spatial frequencies and reconstructing the image from their mean phases and amplitudes. Note that only a half-plane is called for in the Hermitian frequency domain because the image is necessarily real. The direct image accumulation from putting each photon into its pixel is also shown for reference. The magnitudes of the double star 49 Serpentis are both 7.5, their separation is 4 arcsec, and the total number of photons in the observation is 16,384 for somewhat over 1-second total observation time. First notice that Figure 5C compares favorably with the reference in Figure 5A. That means that there were no phase unwrapping errors for any space frequencies of significant amplitude. On the other hand, Figures 5B and 5D are noticeably disturbed. In the

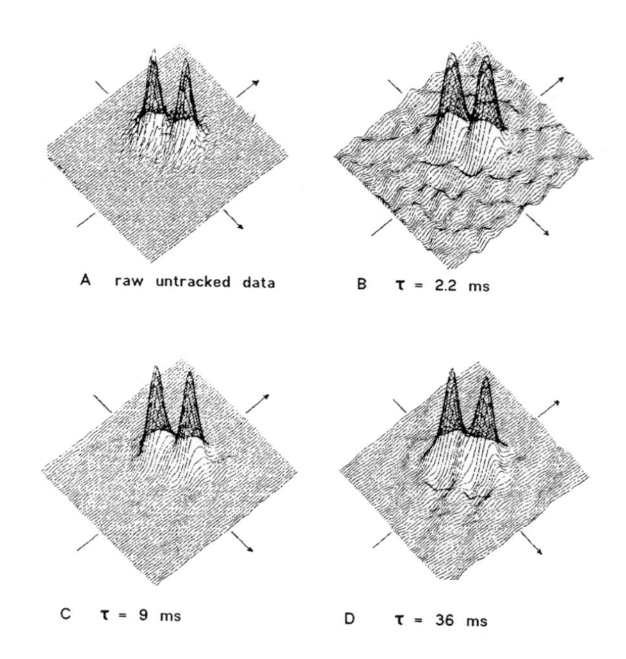


FIGURE 5. Fourier-tracked pictures using various time constants.

former case, there must be too few photons for the temporal bandwidth. In the latter case, the time constant is either too long to keep up with the scintillations or the total observation is too short to reach equilibrium. The efficacity of the tracking is shown in Figure 6, where the peak at about 10 milliseconds is quite consistent with expectations about how short a time is necessary to freeze the seeing fluctuations.

The thing that remains untested is whether the tracking will still work under conditions of speckle agitation. However, in view of the elegant successes of the triple correlation or bispectral methods, whether or not this tracking method works may be largely academic. Due to my lack of working experience with those methods, I will not presume to portray them here.

## Image Stabilization

A simpler, less ambitious goal is partial correction of the image. That for example is the objective of high-speed autotracking mirrors — to stabilize a jittering image. The simplest form depends on a four-quadrant detector for the feedback information. The fault is that that kind of detector senses the median of the image, and the median is dominated by the low space-frequency image components rather than the fine details of the image. It would be far more beneficial to stabilize the details since the low space-frequencies are already, by definition, blurry. Perhaps the most elegant way to stabilize the details is with a correlation tracker.

The operation of a correlation tracker relies on the correlation function between a reference image from the past and the current image frame. The idea is to shift the current image frame to diminish the offset of the peak of the correlation from the zero position. The tracker does not need the entire correlation function, however, and would not be able to use the entire function. An adequate, though nonlinear, measure of the offset is the slope of the correlation function at zero offset.

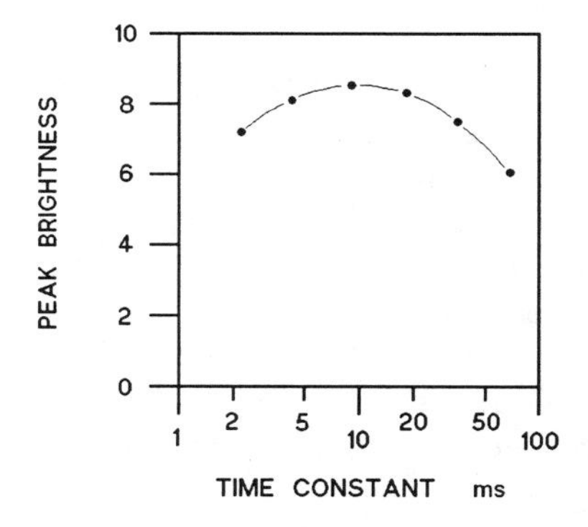


FIGURE 6. Peak star brightness as a function of tracking time constant.

The procedure is quite simple in the image photon-counting mode. We will maintain two variables, initialized to zero: one for the $x$ offset and the other for the $y$ offset. For each incident photon, we add the offsets and then take the differences of the right-left and upper-lower neighboring pixels. The two offsets are additively updated in proportion to the respective differences. Since the image is cumulative, we also have to reduce the factor of proportion with the total number of photons.

With somewhat more difficulty a similar strategy can be followed for CCD detectors. Here we store a reference frame in what is called a dual-port memory. For each pixel of successive frames we take the difference of the products of the incoming pixel intensity with the right and left as well as the upper and lower pixels of the reference frames. The sums of the respective differences for the incoming frame give the $xy$ corrections to be applied, and in this case they are applied to a fast $xy$ steering mirror rather than as numerical shifts. Present technology allows using $32 \times 32$ pixel CCDs at a few hundred frames per second. The results can be impressive for a small patch of sky, called the *isoplanatic patch*. Beyond this patch the corrections would have to be different, and the situation becomes intractable with present technology.

The correlation tracking procedure is essentially the same as the shift-and-add algorithm in speckle imaging. That algorithm involves shifting each frame so as to superimpose the brightest speckle. The brightest speckle is just where the preponderance of high spatial-frequency components happen to be in phase together to make it brightest. It is a very simple algorithm, and it is known to achieve diffraction-limited results although at reduced contrast because the low space-frequency components are not stabilized. For this latter reason the triple correlation or bispectral method has come to be favored even though it is much more computationally intensive.

## Wavefront Tilt Sensors

While tip-tilt wavefront correction by a fast steering mirror works fairly well for small telescopes, it becomes less useful for large telescopes because different portions of the aperture want different tilts. Hence the development of deformable "rubber" mirrors. Two problems remain: the first is to have a reference star within the isoplanatic patch, and the second is to evaluate the wavefront provided by that star. Let me treat the latter first.

Most popular wavefront evaluators are wavefront tilt sensors. Among these is the Hartmann sensor, which disassembles a conjugate image of the telescope aperture into subapertures with a lenslet array. Each subaperture is furnished with a four-quadrant photodetector to notice the star position and hence the local wavefront tilt. Generally there has to be some kind of mathematical reconstruction from tilts to contours for activation of the deformable mirror.

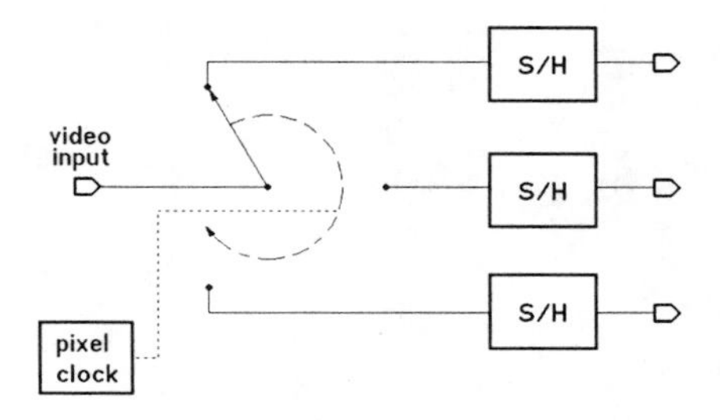


FIGURE 7. Cyclic distributed sample-and-hold.

## Interferometric Wavefront Sensing

For more sensitivity, shearing interferometers also may be used as wavefront tilt sensors. This brings in the additional problems of fringe intensity-to-phase interpretation and two-dimensional phase unwrapping. At least the phase interpretation is easy using the technique from the previous chapter. The reference beam is deliberately tilted to make nominally vertical fringes with a spacing of one fringe per three television pixels. Each successive pixel is distributed cyclically into three sample-and-holds with a commutating switch as shown in Figure 7. Those three values are applied to a hexaflash circuit to get the phase, and the process is easily done in real time, pixel by pixel. The process should be more useful in other applications, such as optical figure evaluation and moiré topography, and a simulated example is shown in Figure 8. On the left are shown the raw television fringes and on the right the corresponding phase displays. The results shown happen to be excessively noisy because of poor circuit-board layout. There are two legitimate objections to the process. The first is that there is an approximation involved when the fringes depart from exactly three pixels per fringe. For a maximum spacing error of a factor of two, the maximum phase error is about 0.1 fringe and for a uniform distribution of spacing errors up to a factor of two, the rms error is about 0.044 fringe. The second objection is that if the object shows reflectance texture, that texture can contaminate the fringe intensities. These objections are of about the same magnitude as objections that apply to other procedures, so simplicity still ought make this the procedure of choice.

An incidental application for this phase measurement procedure is the determination of feature centroids. The features may be either stars or photon-induced events in a picture or lines in a spectrum. Typically we seek a maximum (or minimum) pixel along with its two neighbors, and we want the centroid to be a small fraction of a pixel spacing. The first impulse is find the center of gravity of the pixels values. However, the features then are found to lie systematically close to the pixels. The reason for this is that the contributions from the omitted wings beyond the neighbors of the feature are biased. A much less biased procedure would be to fit a parabola to the three pixels and find its center. Rather than a parabola, our

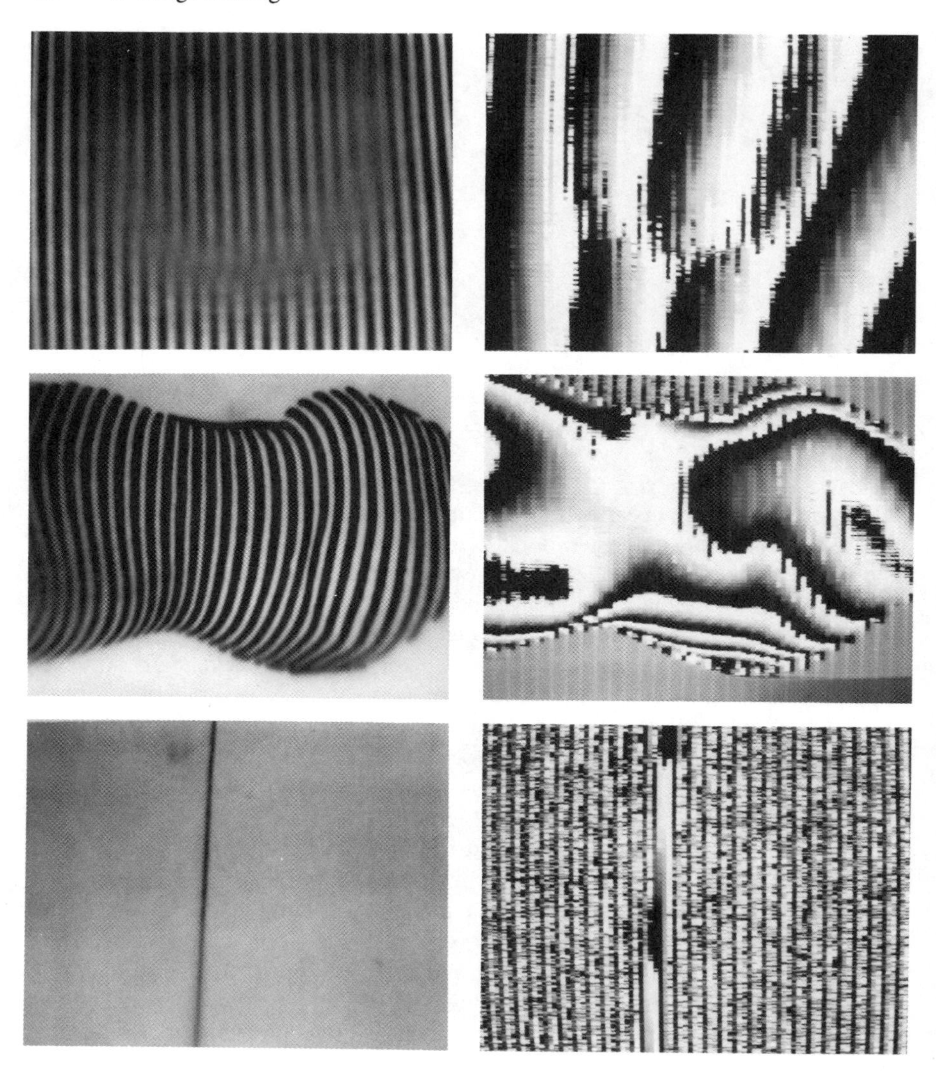

FIGURE 8. Televised interferometric phase display. *Left:* Raw unprocessed display, *Right:* Corresponding phase display.

circuitry finds the phase of a cosine bell that is fit to the pixels. That phase is a trigonometric interpolant that has negligible bias.

## Prismatic Wavefront Tilt Sensor

Returning to the topic of wavefront tilt sensing, phase unwrapping remains a major problem for interferometry. In two dimensions the unwrapping can all too easily wind up with ambiguities resembling Escher staircases. A particularly simple non-interferometric sensor can be based on a triprism. A triprism is a regular tetrahedron that is foreshortened normal to one of its faces. It shows three facets dividing its aperture into 120° sectors and is readily available as a photographic accessory to make triple images. Let us have our telescope focus a star image on the apex of the triprism as shown in Figure 9. A field lens just behind the triprism will form three images of the telescope aperture for three synchronized television cameras to give an RGB (red-green-blue) display. For perfect incidence the display ought to be white, but if any local portion of the aperture is misdirected due to atmospheric turbulence, then that portion will appear as predominantly one color in the RGB display. The results turn out to be almost useless. There is effectively no color blending and black borders appear between the colored regions. When a diffraction-limited spot moves from one facet of the triprism to another, the relevant images do not gradually adjust their intensities accordingly, and while the spot straddles the edge the light gets thoroughly scattered away. Such a situation would give awful servofeedback with its dead zone and bang-bang control.

## Holographic Wavefront Tilt Sensor

What would be called for are gradual edges on the triprism, but that does not work, at least for refractive prisms. However, it can be done with diffractive prisms. Figure 10 shows the scheme. Two coarse orthogonal diffraction gratings function as a holographic double biprism that provide nine pupil images. One is zero-order diffraction, four are $\pm 1$st-order diffraction, and four are cross-product diffraction. Only those four images contained in the dashed outline need be measured, the

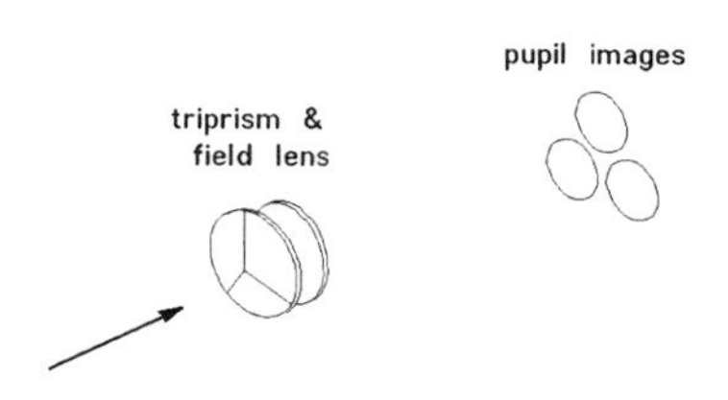


FIGURE 9. Triprism giving pupil images for red-green-blue display.

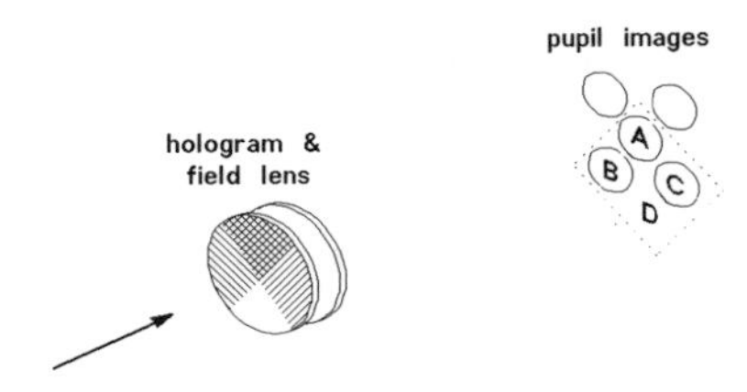


FIGURE 10. Holographic wavefront tilt sensor. The pupil images outside the dashed line are redundant and need not be collected.

remaining images being redundant. The boundaries of the diffraction gratings are not abrupt; the gratings just get gradually weaker across the boundaries. Preferably the gratings are phase gratings with their depths becoming weaker. As a star image moves across the boundary the diffracted pupil images change their relative intensity. The total light will be the weighted sum of the labeled images, $T = A+2B+2C+4D$, where the $D$ corner cross-product images should be weak. The $xy$ deflections then are given by the ratios $\delta x = B/T$ and $\delta y = C/T$. The sensitivity depends on the gradient of the strength of the gratings and need not be linear across the field. The gratings may be fabricated by bleached photography with two exposures (one for each orthogonal grating) or by using photolithography.

Several comments are in order concerning this holographic tilt sensor. The first simply wonders how this sensor performs for speckled star images. The second compares this with Wyant's [1973] double-grating interferometer. The moiré of the double grating is a strength modulation at the difference or beat space-frequency of the two gratings. The modulation is a periodic form of the strength gradient of the holographic device, and behaves like interferometry at reduced sensitivity and reduced need for phase unwrapping.

A third comment relates the holographic device with a theoretical concept. In each of the two dimensions it turns out that if you multiply the image by a linear gradient across the field and compare its power spectrum with that of an unmultiplied image, you can solve for the original image even though the power spectra lack the phases. It functions like a pair of shaded knife-edge tests to evaluate the figure of the wavefront. When the shading stretches across the whole field, however, the system is relatively insensitive, and so the technique has not become adopted. You get more sensitivity but less dynamic range by concentrating the gradient near the center of the field as with the holographic device.

## Wavefront Curvature Sensing

There is a better way than wavefront tilt sensing. The better way, as shown by Roddier [1988], is wavefront curvature sensing. The basic reason that it is a better way lies in its simplicity. All that is needed is to compare a leading-trailing pair of

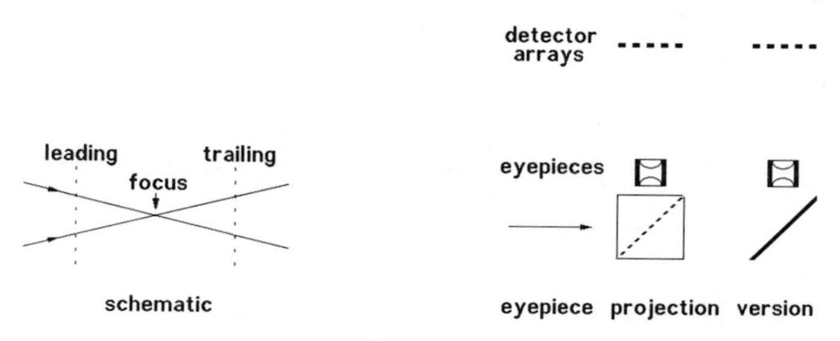


FIGURE 11. Roddier's wavefront curvature sensor.

out-of-focus images as shown in Figure 11. Consider any one of the rays going through the focus. If the regional focus pertaining to that ray is before the nominal focus, then the brightness where the ray intersects the leading image will be more than where it intersects the trailing image. Thus the difference brightness of the images indicates the wavefront curvature. There are boundary conditions that show up as a bright or dark rim on the periphery of the difference image. Roddier points out that Poisson's equation solves the wavefront curvature from the difference image, whose periphery serves as boundary conditions. The system of Figure 11 is remarkably sensitive, even when the spacing between the images is 10 centimeters, so that thermal air currents are easily visible on a television display of the difference image.

The marvelment of curvature sensing comes in its conjunction with what are called *piezo bimorph deformable mirrors*. When a thin piezo disc that has been polarized is subjected to a voltage, not only does its thickness change, but its diameter also changes. A lamination of two discs arranged so that one dilates while the other shrinks then will buckle according to the voltage. These bimorphs are readily available because they are used ubiquitously as compact inexpensive loudspeakers. Often one of the laminations is a passive material, such as brass. The deformation-versus-voltage responsivity of these devices goes as the square of their aspect ratio, so it is very desirable to have very thin layers. One need only replicate a mirror off of an optical flat onto the passive layer and etch the electrode on the piezo layer into island electrodes, as is shown in Figure 12, to convert the loudspeaker to a useful deformable mirror. The reason that this is so delightful for curvature sensing is that this deformable mirror effectively solves Poisson's

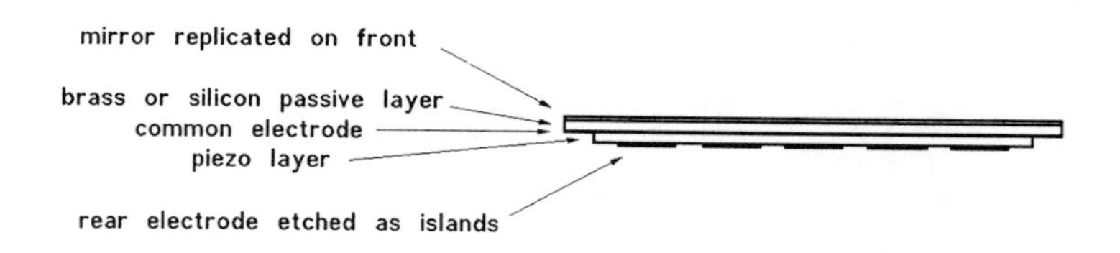


FIGURE 12. Bimorph deformable mirror construction.

equation. Thus we can directly connect the pixel detectors of the curvature sensor to the island electrodes of the deformable mirror to get proper adaptive optics. There is no need for intervening computation or phase unwrapping.

The bimorph mirror is very light and should be mounted on a three-point support, two of which are piezo stack actuators for overall tilt control. The information for this tilt control comes from pixel detectors at the boundary of the pupil. Also, it might be preferable to choose a silicon wafer, which is quite readily available from the semiconductor industry, rather than brass as the passive layer.

## Prismatic Variants

With a single array of detectors and optically switching between the two pupil images, both the geometry and the responsivity of the array would be automatically the same for both images. The one concern is that the optical switching must include a 180° rotation for proper mapping of the rays. This could be accomplished using the composite prism illustrated in Figure 13. The beamsplitters are polarization-dependent, and switching would be done using a liquid crystal or a photoelastic polarization switcher. One polarization takes the short route while the other takes the long route through the composite prism. In the long run, however, this prism approach is more intricate and probably would be more expensive than making two geometrically matched detector arrays and using the configuration of Figure 11. The relatively large size of the discrete detector arrays calls for a pair of eyepieces to project the two out-of-focus image planes to larger dimensions as was shown in Figure 11.

Nevertheless, the prism of Figure 13 does give occasion to mention an elaborated configuration shown in Figure 14. With careful adjustment the two routes through this prism can be the same length, while still giving the 180° difference of image rotation. Thus it can act as a white-light 180° shearing interferometer. Ordinarily, such an interferometer is made as a Michelson with either a cube corner at the end of one arm or roof mirrors at the ends of both arms. The problem with the

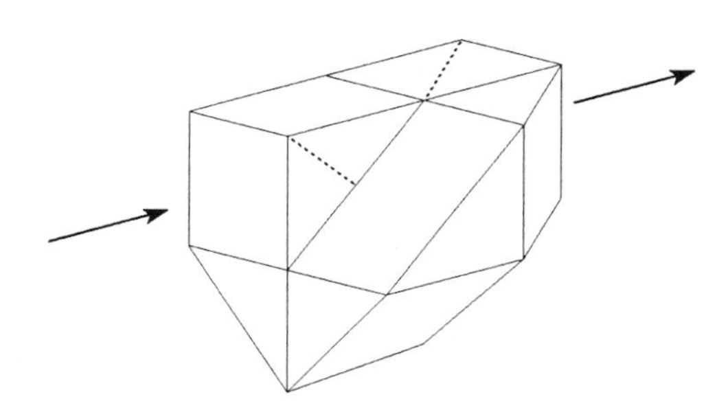

FIGURE 13. Prism for wavefront-curvature sensor.

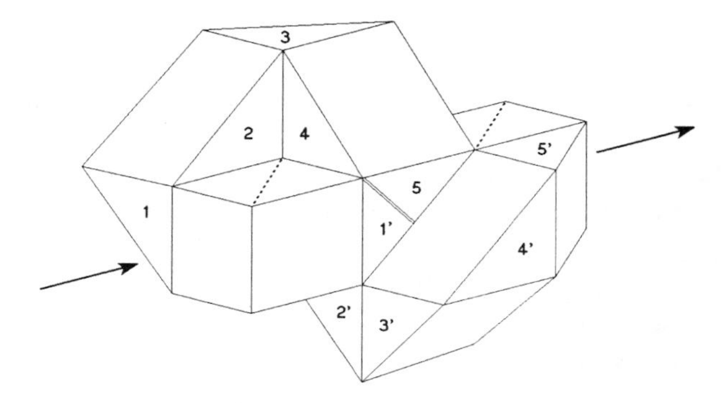


FIGURE 14. 180° rotational shearing interferometer prism. Numbers indicate prism sequences with primes giving image inversion.

ordinary configuration is that the cube corner or roof angles must be fabricated with exquisite precision. The advantage of Figure 14 is that no such precision is required, and there is also the cosmetic feature of straight-through viewing. On the other hand, the Michelson with roof mirrors has the definite advantage that it can be adjusted for rotation angles other than 180°, and that can be helpful when mathematically converting from interferograms to images. The underlying reason is that phase closure techniques that can overcome phase perturbations require other than 180° rotation, but that is beyond my scope here.

The major problem with these adaptive optics techniques is the requirement for a reasonably bright source within the isoplanatic patch. Commonly in astronomy we would like the wavefront correction in order to perceive very faint sources, and the chances of finding a bright source within a very few arcseconds are slight. One very daring technique that has been implemented is to create an artificial source by shining a high- power laser into the upper atmosphere. The scattering, and it can be either Raman or resonant off of sodium atoms, creates the artificial reference star. This technique turns out to work far better than I would have expected, considering the apparantly large size of the artificial star. However, it is not a technique that is likely to be adopted by many observatories because of the potential dangers and regulations involved.

Although speculative, I am led to wonder whether passive observation of the atmosphere might suffice. Consider the thermal emission on the edge of an atmospheric absorption band. If the upper atmsophere appears mottled, in much the same way that stellar illumination on telescope objectives appears mottled, then that pattern could serve as information leading to wavefront correction. It is a question of whether infrared detector arrays for the 8- to 12-$\mu$m region are adequately sensitive to perceive mottling, and whether that mottling would properly pertain to the atmospheric level responsible for perturbing the astronomical images.

# Bibliography

P. A. Ekstrom, 1981 "Triggered avalanche detection of optical photons" *J. Appl. Phys* 52: 6974–6978

A. Labeyrie, 1970 "Attainment of diffraction limited resolution in large telescopes by Fourier analysing speckle patterns in star images" *Astron. and Astrophys.* 6: 85–87

M. Lampton and F. Paresce, 1974 "The RANICON detector" *Rev. Sci. Instr.* 45: 1098

B. L. McGlamery, 1970 "Image restoration techniques applied to astronomical photography" in *Astronomical Use of Television-Type Image Sensors*, V. R. Boscarino, ed. NASA Publication SP-256

L. Mertz, 1979 " Speckle imaging, photon by photon" *Appl. Opt.* 18: 611–614
———1985 "Multichannel seeing compensation via software" *Appl. Opt.* 24: 2898–2902
———1989 "Hexflash phase analysis examples" *Proc. SPIE* 1163: 39–43
———1990 "Prism configurations for wavefront sensing" *Appl. Opt.* 29: 3573–3574

L. Mertz, T. D. Tarbell, A. Title, 1982 "Low noise imaging photon counter for astronomy" *Appl. Opt.* 21: 628

C. Papaliolios and L. Mertz, 1982 "New two-dimensional photon camera" *Proc. SPIE* 331: 360–364

F. Roddier, 1988 "Curvature sensing and compensation: a new concept in adaptive optics" *Appl. Opt.* 27: 1223–1225

J. G. Timothy and R. L. Bybee, 1977 "Multi-anode microchannel arrays" *Proc. SPIE* 116: 24

J. C. Wyant, 1973 "Double frequency grating lateral shear interferometer" *Appl. Opt.* 12: 2057

# 5

# Spectroscopy

Spectroscopy is a major tool for learning the nature of celestial sources. It tells us about composition, temperatures, motions, and the general physical circumstances of those sources. The most common instrument for spectroscopy is a diffraction-grating spectrograph, which spreads the light into its panoply of colors. For many years spectrographs were designed to use photographic plates, but now CCD detectors are used because they are more sensitive and give photometrically better results more directly, without the need for wet processing or densitometers. The optical design of the spectrographs nevertheless remains much the same, but with gains accruing from improvements in the quality of the optics. There are, however, several designs that have been quite overlooked.

## Following Dyson

In 1956, Dyson presented a remarkable design for a unit- magnification optical system that has no Seidel aberrations for any f-number. It consists of a hemispheric lens and a concentric spherical mirror. The use of only spherical surface figures renders the configuration to be eminently practical as well as elegant. The design converts easily to a spectrograph as shown in Figure 1 by making the mirror a concave diffraction grating.

This arrangement has a number of virtues, which are listed below. There is some redundancy in the list in order to emphasize certain aspects.

1. Sharp imagery due to the inherent absence of Seidel aberrations.
2. High numerical aperture, with N.A.>0.6 being practicable even though it gives very small depth of field to make focusing difficult.
3. Stigmatic, which is very important for energy concentration.
4. Flat field, which is important for flat detectors.
5. Wide unvignetted field having linear dispersion as a function of wavelength, covering the complete photographic spectral range and permitting long slits.
6. Readily accessible field located on the exterior of the spectrograph.

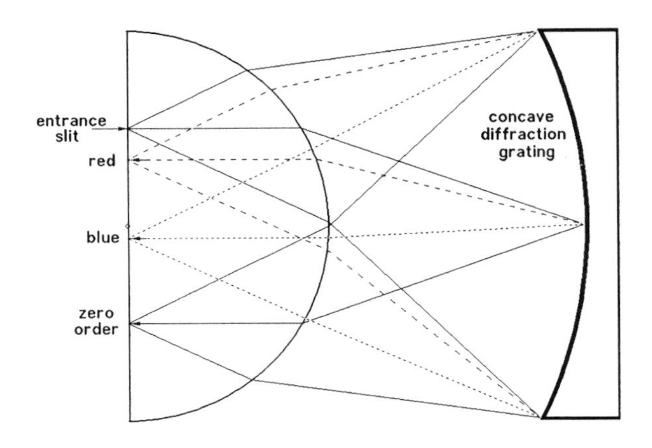


FIGURE 1. Dyson's unit-magnification optical system as a spectrograph.

7. Nonanamorphic field having equal magnification along and across the dispersion.
8. Telecentric; with its pupil situated at infinity, focusing errors introduce neither dispersion change nor aysmmetric intrument profiles.
9. No central obscuration of the pupil.
10. No aspherical surfaces are required.

The following list of restrictions is relatively modest:

1. Limited to low dispersions.
2. Transmitting material is required.
3. Focal ratio does not directly match that of telescopes. This can be cured with a microscope objective focal reducer. (The focal ratio does match that of fibers.)
4. The grating must be holographically formed, and so it may be more difficult to achieve efficient blazing.

The extraordinary optical performance of the Dyson configuration can be accounted for in terms of concentric retroreflectivity. The fact that it is concentric means that there can be no skew rays. Rocking the system about the axis that connects the entrance to the zeroth-order exit image changes nothing. Consequently, the sagittal focusing is necessarily ideal, without aberrations. If the system is also a retroreflector, then an entering ray will exit parallel and at an equidistance opposite the center of concentricity. Hence, rocking the system normal to the previous axis (in-plane rotation of the figure) would change nothing and tangential focusing would be ideal. However, because of the spherical aberration of the hemispheric lens, the retroreflectivity only pertains to one zone. The second rocking thus reduces the off-center distance of the entrance ray by the cosine of the rock angle so that the sytem is no longer perfectly retroreflective, and so there is a second-order departure from perfect tangential focusing. Since we do want the off-center zone of the entrance slit to be retroreflecting, that means the spherical lens should be have a slightly larger radius than the paraxial prescription given by Dyson.

Wynne [1969] has suggested broadening the satisfaction of the retroreflective condition while maintaining concentricity by making the hemispheric lens a monocentric doublet. The shell of lower index glass reduces the spherical aberration of the doublet as compared with a singlet. Shafer [1991] has carried it even further by pointing out that continual gradation of the index from the center all the way to the reflective mirror, so that the lens becomes half of a Luneberg sphere, gives ideal correction. His result is a perfect imaging system with no geometric aberrations of any order.

The validity of the image quality for the diffracted spectra has been established experimentally from several prototypes rather than by numerical ray tracing. The initial prototypes used ordinary condenser lenses which are less than hemispherical, but they only omit a plane parallel portion of glass and so contribute some spherical aberration to the imagery. Nevertheless, the image quality was extraordinary for an $f/1$ spectrograph with a 7.5-centimeter grating diameter at 200 Angstroms per millimeter dispersion.

I learned a number of things from the several experiments. First was that the image quality, known to apply to the undiffracted light, also applied to the diffracted spectra. In particular, the spectra were stigmatic, flat-field, and showed linear dispersion with wavelength. There was some distortion in that the spectrum lines appeared slightly curved in the field.

Second was that the paraxial recipe was improper. What really counts is that the zone of the entrance slit be retroreflective. In other words, for that zone the spherical aberration of the hemispheric lens would focus parallel entrance light exactly onto the surface of the concave grating.

The third had to do with making the grating holographically in situ. A beamsplitter arrangement as shown in Figure 2 could be used to make two coherent point images from a blue (HeCd) laser. The spacing is used to define the desired

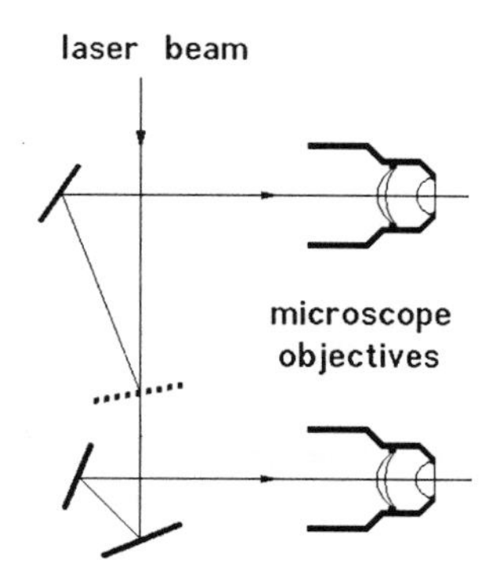


FIGURE 2. Making two coherent point images.

dispersion of the spectrograph. The two point images should be arranged in the field of view such that both points lie in the good zone. This means that the points span a chord in that zone, rather than lying on a diameter in the field.

Incidental things I learned were that the concentric alignment could be easily done with a point light source emanating from an optical fiber by simply superimposing the (zeroth-order) image from the grating with dielectric reflection from the condenser lens, and both with the fiber source. I also learned to appreciate that the depth of field associated with fast $f/0.7$ optical systems is so small that it makes focusing quite difficult.

It is instructive to compare the Dyson arrangement with the traditional Rowland configuration and aberration-corrected holographic gratings. Figure 3 illustrates Rowland's circle, the locus of the tangential foci. It also works with the undiffracted light, so that the intersections of the circle with any vertical line give unit-magnification conjugates of the concave mirror for the tangential foci. You may note that, for any pair of intersections, perfect imagery would require an elliptical mirror with those intersections as foci. Hence the Rowland circle is just a first-order locus of spherically aberrant foci and will not be good with fast gratings. Unit-magnification conjugates for the sagittal foci lie on the plane S, so there is lots of astigmatism as soon as you get off-axis with a spherical mirror. The diffraction grating can be made to correct holographically for the astigmatism at specific diffraction angles. The locus of astigmatically correct images is shown as the curious curve in Figure 3. Dyson's hemispheric field lens flattens out the

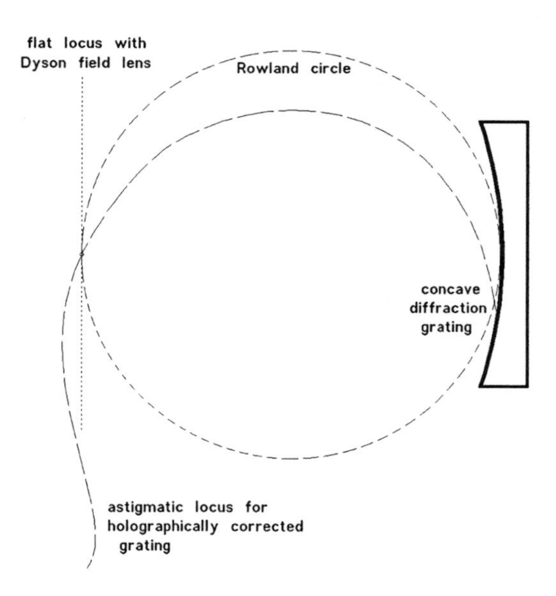


FIGURE 3. Rowland circle and other focal loci for a concave diffraction grating.

Rowland circle to coincide with the sagittal plane and thus correct the astigmatism over that entire plane.

An unusual application of the Dyson configuration would be as a coded-aperture spectrograph. In this case, the entrance slit is replaced by a coded aperture, which may be a Fresnel zone pattern, so that the dispersed image is the convolution of the aperture pattern with the spectrum. The coding will be beneficial only under certain conditions. First, the source would have to be diffuse in order to fill the coded aperture. Galaxies will fulfill this condition if their integrated spectrum is sufficient, but not necessarily if the variations of the spectrum for different regions of the galaxy are sought. Even stars may fulfill the condition for a large telescope where the atmospheric turbulence blurs the image. A second condition is that Fellgett's multiplex advantage be applicable for the detector. Generally this means that the detector noise must not rise any faster than the square root of the illumination. The exception is emission line spectra, when only the main emission lines rather than the background between them are of interest. Photographic detectors, with their threshold and toe on their density-exposure curve, fulfill Fellgett's advantage. CCD detectors also fulfill his advantage insofar as their preamplifier noise is concerned. Photon-counting detectors do not fulfill his advantage.

## Following Offner

Offner [1975] invented an all-reflecting version of the Dyson configuration that should be similarly suitable as a spectrograph as shown in Figure 4. This case requires a convex diffraction grating and so the spectrograph has not been tested. Nevertheless, it offers an interesting perspective that allows for further development. In particular, we may note that Offner's invention is a Schmidt telescope, where the pupils are interchanged with the image locations. Accordingly, we note

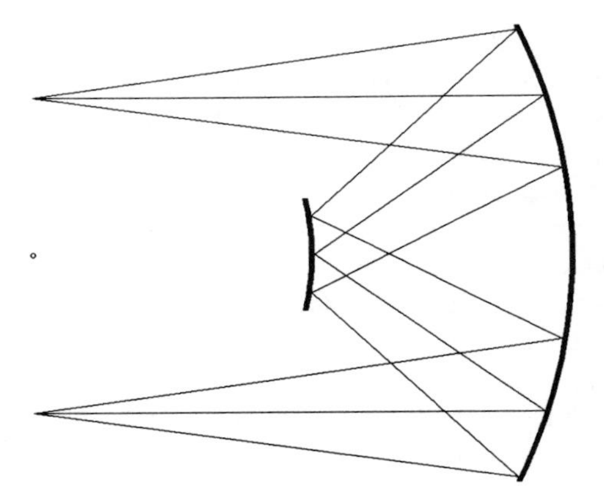

FIGURE 4. Offner's unit-magnification optical system.

that in principle there ought to be a Schmidt corrector plate as a field lens at the focal plane, and that the proper location for the entrance slit would lie in the neutral zone of that corrector. Its neutrality there allows us to omit the corrector for the full field of the spectrograph in practice even though the neutral zone is annular, the only penalty being a slight sacrifice of telecentricity in the other portions of the field.

The analogy with a Schmidt telescope invites changing to a Maksutov-Bouwers telescope or a monocentric Mangin reflector. These configurations correct the spherical aberration of the concave mirror with a meniscus rather than the Schmidt plate, and so fully satisfy the concentricity and retroreflectivity conditions. Hence they are less restricted by the width of a neutral zone and provide ideal imagery over a broader field than Offner's. These revisions were devised by Schafer when he found that placing the meniscus on the convex mirror did give that much broader field. The need for a transparent meniscus unfortunately rules out application toward photolithograpy in the soft X-ray region.

## Whispering Galleries

Getting back to the Offner-Schmidt analogy, some tantalizing things happen when you move the neutral zone farther and farther from the center. Concomitantly, the radius of the convex grating increases so that it intersects the axis where the spherical aberration of the concave mirror causes parallel light to focus on that axis. When the zone offset becomes sin 45° times that of the concave mirror, so does the radius of the concave grating, and we find grazing incidence on the grating. If we now omit the grating, we still find a good undiffracted image opposite the entrance slit. This is a low-order whispering mode having two reflections off the concave hemispheric mirror. The configuration now can be folded with a flat mirror so that the good image coincides with the entrance. Furthermore, we can make that flat mirror as an ordinary plane-diffraction grating to obtain the configuration shown in Figure 5. The "Rowland circle" of the concave mirror is sketched in to show that the entrance is imaged onto the grating, so that it would be useless for the grooves of the grating to be oriented normal to the paper in Figure 5A. That image is extremely astigmatic, however, and the system works quite well with the grooves oriented parallel to the paper. That astigmatism on the grating is such that gratings with a very elongated aspect ratio are quite suitable, even when the original entrance pupil is circular. Figure 5B shows the system looking down the entrance ray so that the image plane is the plane of the paper. While the dispersed spectrum is stigmatic and lies in a plane, it lies as shown on an arc in that plane. That may be more inconvenient than a line, but it does allow spectral scanning by simply rotating the grating. Since plane gratings and spherically figured mirrors are so readily available, I have been able to verify the configuration experimentally with even a very crude setup. The annular zone suitable for the entrance is rather narrow, ruling out long entrance slits, but the arrangement might be suitable for wavelength-division multiplexing in fiber-optic communication. Another application might be

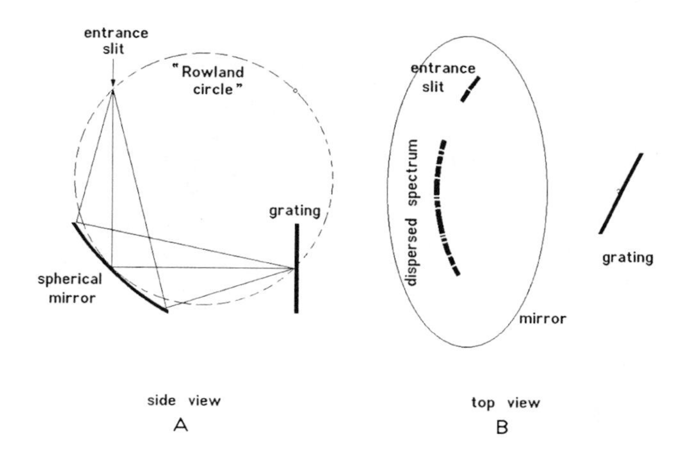


FIGURE 5. Spectrograph using concave mirror.

as an echelle spectrograph. It should not be too difficult to incorporate a weak dispersion between the entrance and the spherical mirror, and even to include at the entrance a cylindrical field lens that would approximate that morsel of a Schmidt corrector plate.

There is one mild aberration along the dispersion in the configuration of Figure 5. Those rays that impinge upward or downward onto the grating in Figure 5A experience a foreshortened delay between the grooves as compared to the principal ray. As a result, there becomes a blue-shifted wing on the instrument profile. That wing might be cured if the morsel field lens were redesigned slightly away from the plane of the entrance slit. In that fashion the wing could be displaced back to its proper position, although the correction would only apply strictly at one wavelength in the dispersed spectrum.

Further things happen when we move the neutral zone out beyond sin $45^{c}irc$ times the concave radius, again moving us away from grazing incidence on the grating. The grating curvature flips from convex to concave because we approach it from its other side, and when we get to sin 60° times the mirror radius the grating becomes conformal with that concave mirror as shown in Figure 6. Now we have rather steep incidence on the grating so as to get high dispersion. The incidence is also steep enough ($> 45°$) so that even symmetrical groove profiles can give efficient blazing. Symmetrical groove profiles are relatively easy to obtain with holographic fabrication of the grating.

This configuration might be applied in several ways. You get maximum dispersion by fabricating the grating using two coherent sources as from Figure 2 that span the diameter of the neutral zone. Littrow operation with the spectrum arriving back near the entrance slit then allows us to dispense with the lower portion of the concave mirror after fabrication of the grating so that the spectrograph uses only one concave mirror along with the concentric concave grating. Alternatively,

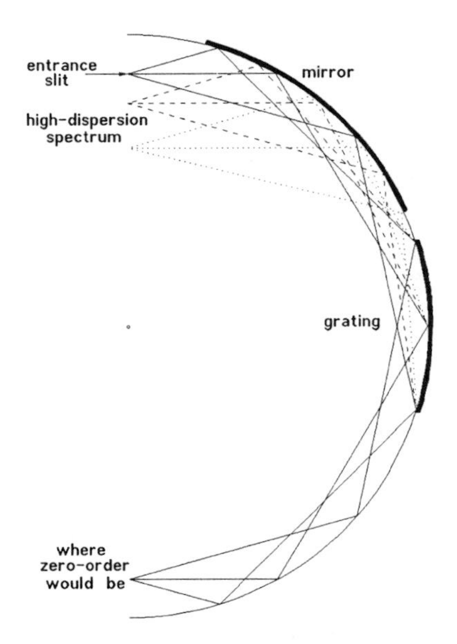


FIGURE 6. Whispering-gallery spectrograph.

we could leave in the lower portion of the mirror to get the very-short-wavelength portion of the spectrum near the zeroth-order image. The steep incidence aids the reflectivity at short wavelengths.

All of these concentric spectrograph configurations will give unexcelled image clarity and are inexpensive to make because of their strictly spherical figures. However, since they have not been described in textbooks, they have not been adopted, and it is a bit of a shame to see the very complicated lenses and Schmidt arrangements that are actually built.

## Fourier Transform Spectroscopy

Turning from dispersive spectroscopy to Fourier transform spectroscopy we find a technology that has matured. Only a brief synopsis of the technique is included here to provide a context for some incidental comments. Fellgett [1951 Cambridge U. Thesis] pointed out that a scanning interferometer serves to modulate the intensity of a light beam such that the modulation frequencies are in proportion to the optical frequencies. Even more importantly, he pointed out that in many circumstances there is a signal-to-noise advantage to be gained by measuring the modulated intensity and analyzing its modulation frequencies to get the optical spectrum, as compared to measuring the spectrum one color at a time. Another advantage that remains important is the throughput, or Jacquinot, advantage whereby the absence of an entrance slit allows the interferometer to accept all of the light from a large telescope, even when that light is blurred by seeing.

The modulated intensity is called an *interferogram,* and there is a point on that interferogram known as the *white-light fringe* where all of the modulation frequencies are in phase. If there is a sample at that point and the phase at that point is either 0° or 180°, then the interferogram is perfectly symmetrical and a Fourier cosine transformation suffices to derive the spectrum. The symmetry also implies that the two sides of the interferogram are redundant, so that only one side need be recorded. The interferogram usually is nonsymmetric, however, so exactly where to choose the center for the Fourier transform is slightly ambiguous. The on-center phase may be other than 0° or 180°, and the central position may even be a function of wavelength. The latter situation is refered to as *chirped.* Mathematical compensation for the lack of symmetry is called *phase correction.* The purpose is to get the spectrum that would have been obtained had the interferogram been perfectly symmetrical, and also permit the use of severely off-center interferograms.

Correction recipes may be in either the time (interferogram) domain or in the spectral domain, but their use is often rote or poorly understood and computationally inefficient. All of the prevailing recipes use a very short piece of the interferogram around the white-light fringes to characterize the phases and then the entire interferogram to extract the detailed spectrum. The two prevailing methods (correlational and multiplicative) of phase correction are related by the following mathematical statement. The real part of the Fourier transform of the cross-correlation of two real functions is equal to the scalar product of the Fourier transforms of those two functions:

$$\mathrm{Re}\{\mathcal{F}(I * K)\} = \mathcal{F}(I) \cdot \mathcal{F}(K) ,$$

where $I$ and $K$ are truncated versions of the interferogram. Figure 7 gives a graphic depiction of the off-center interferogram along with a short reference truncation function and a long truncation, so that $I = T_L I_\infty$ and $K = T_R I_\infty$, respectively. The long truncation $T_L$ has a step so that its symmetric part about the white-light fringe will be a clean boxcar. Then we also divide both sides of the equation by the modulus $|\mathcal{F}(K)|$:

$$\frac{\mathrm{Re}\{\mathcal{F}(I * K)\}}{|\mathcal{F}(K)|} = \frac{\mathcal{F}(I) \cdot \mathcal{F}(K)}{|\mathcal{F}(K)|} .$$

The left side represents the correlational form of phase correction, while the right side represents the multiplicative form. Note that the numerator on the left will be practically symmetric around the center since it is practically an autocorrelation. If we now choose the identity $I = K+(I-K)$ and note that $\mathcal{F}(K)\cdot\mathcal{F}(K)/|\mathcal{F}(K)| = |\mathcal{F}(K)|$, we can expand the equation to

$$|\mathcal{F}(K)| + \frac{\mathrm{Re}\{\mathcal{F}\big((I-K) * K\big)\}}{|\mathcal{F}(K)|} = |\mathcal{F}(K)| + \frac{\mathcal{F}(I-K) \cdot \mathcal{F}(K)}{|\mathcal{F}(K)|} .$$

Although it may seem longer algebraically, this expression is easier computationally than prevailing techniques. The use of the truncating function $(T_L - T_R)$

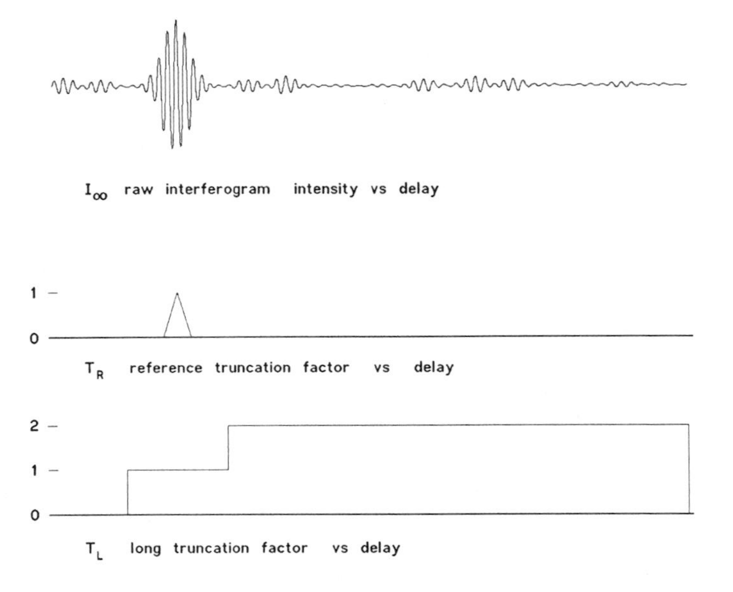


FIGURE 7. Truncation factors for phase correction in Fourier transform spectrometry.

eliminates the strong white-light fringes to reduce dynamic range problems and detector nonlinearities in the computation of the long Fourier transform. Furthermore, we see that phase correction simply adds details to the modulus $|\mathcal{F}(K)|$ of the reference spectrum. The choice of $T_R$ as a triangle function ensures that there are no negative side lobes to upset that modulus, and the shortness of $T_R$ minimizes the positive contamination from the nonlinearity of the modulus function. $T_R$ should be just long enough to characterize the phases clearly.

The computation is carried out most easily by using the long truncation as the real part and the short reference truncation as the imaginary part of a complex Fast Fourier Transformation (FFT). There should be no relative shift between the truncations even though the short one includes mostly zeros. Furthermore, both truncations should be padded with zeros to be at least as long as the symmetric length of the long truncation function. That padding ensures that the FFT yields the spectral resolution commensurate with a two-sided interferogram. At the conclusion of the FFT, the two spectra are separated as the symmetric and antisymmetric parts of the complex result, which extends up to the sampling frequency, i.e., twice the Nyquist frequency. It is then straightforward to calculate the right-hand side of the last phase correction equation. There is no need to explicitly calculate phases.

However, if you should calculate the phase spectrum and unwrap it as a function of frequency (according to the prescriptions given in Chapter 3), then you can expect to find a fairly linear function where there is reasonable signal-to-noise. That function will be steep as long as the white-light fringe is offset from the origin of calculation. Subtracting the linear gradient is equivalent to shifting the reference origin to that fringe. If the extrapolated intercept $\phi_0$ of the phase axis happens to be $0°$ (or some multiple of $360°$), then the interferogram was perfectly symmetric with a bright central fringe. If $\phi_0 = 180°$, then the symmetric interferogram had a dark central fringe. If there is any curvature to the phase spectrum, so that $\phi_0$ depends on frequency, then the interferogram is said to be chirped. There are occasional pathological cases that may occur in the infrared, where for some spectral regions the source acts as a sink of radiation, leading to the $180°$ jumps in the phase spectrum. These may be best dealt with interactively during the phase correction process.

## Nonlinear Analyses

For the most part, Fourier transformation is accepted as the exclusive way to arrive at the spectrum on account of the mathematical rigor. Although that would be the case for infinite interferograms, in practice the Fourier transform serves only to estimate the spectrum. The quality of that estimation tends to be quite good in those cases where the interferogram has decayed completely away before the cessation of measurement. Otherwise, the estimate is faulty in that the spectrum is poorly resolved and the instrumental profile shows its side lobes. The implicit extrapolation that the interferogram would have been zero beyond the termination of measurements loses its credibility and opens the way to more modern spectral estimation. For example, autoregressive estimators characterize successive samples of the interferogram as linear combinations of prior samples and thus extrapolate the interferogram. The maximum entropy method turned out to be equivalent to the autoregressive method, but there are also several others such as maximum likelihood and Prony's method. The descriptions and details of these methods can be found in the books by Kay [1988] and Marple [1987]. While most of the emphasis of these modern spectral estimators has been on improving the resolution, other benefits are more important for Fourier spectrometry. After all, improved resolution already is available simply by extending the scan of the interferometer.

One situation is emission line spectroscopy in the optical or ultraviolet regions, where we might want to detect faint emission lines in the neighborhood of bright ones. The applicability of Jacquinot's throughput advantage for interferometers (the absence of entrance slit) and Connes' precision advantage (digital frequency comparison) encourage Fourier spectroscopy. Although Fellgett's multiplex advantage does accrue for the bright lines, it becomes a distinct disadvantage in the dark regions between the lines, masking the faint lines. That is because the photon shot noise from the bright lines gets spread out uniformly, spilling into the dark regions. Note that the presence of bright lines in the spectrum intimates that the interferogram has not decayed away and might be usefully extrapolated. The situation is the same as that encountered in Chapter 2 for indirect imaging of X-ray stars. As there, the implicit nonnegativity of modern spectral estimators can serve to suppress the noise in the dark in-between regions. Also, as with the

X-ray situation, there might be a rationale in minimum entropy criterion, whereby the spectrum is sought to consist of relatively few discrete lines. Inasmuch as the modern spectral estimators are nonlinear, the Fellgett multiplex signal-to-noise relationships that we had previously thought to be fundamental have lost their applicability.

Another utility for modern spectral estimation would be in the determination of continuum levels in the spectroscopy of absorption lines. An accurate knowledge of the continuum level is necessary for quantitative evaluation of equivalent widths. Using the low-resolution spectrum from the reference interferogram is not at all suitable because that spectrum gives an average level that is systematically lower than the continuum for absorption line spectra. Suppose, however, that we were to backextrapolate the interferogram from the high retardations back into the region of the reference interferogram, and then subtract the extrapolation from the actual interferogram. The extrapolation would be based on autoregression (maximum entropy) so that the subtrahend would represent Lorentzian spectral line contributions to the interferogram. The remaining interferogram then will correspond to the contributions from the continuum. While this process still calls from some judgement as to where the continuum interferogram has decayed away, it still seems to me to be more objective than the usual procedure of trying to estimate a smooth envelope over the calculated spectrum.

Finally, we may ask ourselves the question of whether the spectrum is our goal or whether the spectrum is just a step toward some further physical understanding? Do the concepts of automatic control pertain to atomic and molecular spectroscopy? For automatic control we are concerned with the response and stability of systems and filters with various excitations. The interferogram, for example, may be considered as somewhat of an impulse response relating to the source, even though it lacks the property of causality. Under Fourier transformation we evaluate the coefficient series of the polynomial in $\exp(i\theta)$. For Prony's method we consider the interferogram to be a rational function in $\exp(i\theta)$, having a polynomial in a denominator as well as in the numerator. For the maximum entropy method, the numerator polynomial is taken to be unity. Common practice in control theory is to factor the polynomials. The roots of the numerator polynomial are called roots, while those of the denominator polynomial are called poles. Hence the maximum entropy estimation is said to be an *all-pole model.* In control theory, the locations of the roots and poles in the complex plane, whether they lie within or outside the unit circle and in the right or left half-planes, are critical to the understanding and stability of the system. The question becomes whether the same might apply for spectroscopy. At least it might serve to condense redundancy from a host of spectral line information, much as the Rydberg expression condenses many lines from the hydrogen spectrum, leading to an improved physical understanding in terms of atomic energy levels. It is not at all clear whether a comparable improvement might accrue from the root-pole analysis of interferograms.

# Bibliography


J. Dyson, 1959 "Unit-magnification optical system without Seidel aberrations" *J. Opt. Soc. Am.* 49: 713

S. M. Kay, 1988 *Modern Spectral Estimation,* Prentice-Hall

S. L. Marple, 1987 *Digital Spectral Analysis,* Prentice-Hall

L. Mertz, 1977 "Concentric spectrographs" *Appl. Opt.* 16: 3122
———1991 "Concentric systems for adaptation as spectrographs" *Proc. SPIE* 1354: 457–459

A. Offner, 1975 "New concepts in projection mask aligners" *Opt. Eng.* 14: 131

D. Shafer, 1991 "New perfect optical instrument" OSA Annual Meeting paper MN1, *Technical Digest* p. 13

C. G. Wynne, 1969 " A unit-power telescope for projection copying" in *Optical Instruments and Techniques 1969* ed. J. Home Dickson, Oriel Press

# 6

# Pulsars

## Introduction

Pulsars are always said to be rotating neutron stars. The grounds for that belief were launched very quickly after pulsars were discovered. The regularity of the pulses suggested that either pulsation or rotation must serve as the clock, and the fast rate of the pulses then requires an extreme density, since for gravitationally bound objects the frequency is limited by the square root of the density. The object would simply fly apart at higher frequencies. Even the densest known objects, the white dwarf stars, could not account for the fast pulsars, and so astronomers turned to the even more dense theoretically anticipated neutron stars. Since the vibration frequency of neutron stars was too high, the conclusion settled on rotating neutron stars.

That conclusion quickly became reinforced with corroborating evidence. A fast pulsar was discovered in the Crab nebula that had historically appeared as a supernova in 1054 A.D. A neutron star could be expected to form and remain after a supernova explosion. Furthermore, the Crab pulsar (as well as all other pulsars) turned out to be slowing down slightly, and the rotational kinetic energy from the slowdown was found to suffice to sustain the high-energy electrons of the nebula. For lack of alternatives the evidence appeared convincing even though the radio emission mechanism remains equivocal to this day.

## Cavity Clocking

Well there is an alternative. The clock need not involve a gravitational restoring force, but it can be governed by an electromagnetic cavity much like the Schumann resonances. The Schumann resonances are the around-the-world modes of the shell cavity lying between the earth's surface and the ionosphere, and are at approximate multiples of $7\frac{1}{2}$ Hz. Note that this is a bona fide astronomical cavity having a characteristic period corresponding to a fairly fast pulsar.

Before going into the technical discussion, let me briefly narrate the inception of the idea. One Thursday morning in 1968, shortly after the announcement of pulsars, D.J. Bradley gave a lecture at Harvard Observatory on the topic of mode-locked lasers. The topic was quite modern at that time. Among his slides was one showing the light intensity from a mode-locked laser as a function of time. That very afternoon, Peter Strittmatter gave a colloquium on the discovery of pulsars. Among his slides was a chart record of the radio pulses from CP1919, the first discovered pulsar. I found the resemblance of the slides to be uncanny, and was thereby stimulated to learn about mode-locking and to seek a suitable cavity. What I found was that mode-locking is such a simple phenomenon as to be easily expected, and that cavities with just the right characteristic periods should exist at the surfaces of ordinary degenerate dwarfs.

## Degenerate Dwarfs

The trait that defines degenerate dwarfs is that, unlike ordinary stars, they are not supported against collapse by nuclear energy providing thermal and radiation pressure; they are supported by electron degeneracy pressure. They are much smaller and denser than ordinary stars. They still have a thin nondegenerate surface layer of hydrogen, but going inward from the surface the hydrogen rapidly gets compressed to the metallic phase by the strong gravity. The metallic phase is roughly twice as dense as the dielectric phase of hydrogen, so that the strong gravity settles it as a metallic sphere having a sharp boundary, much as the surface of a lake. The shiny surface of that sphere provides the lower reflecting boundary of our cavity. Above the sphere lies dielectric hydrogen in various degrees of compression. It will be the gradient of the refractive index of that dielectric hydrogen that will serve to refract rays downward and act as the upper boundary of the cavity. This proposed configuration is shown in Figure 1.

The condition that rays will spiral inward to effect closure of the cavity is

$$\frac{d(2\pi nr/c)}{dr} < 0 \qquad \text{or} \qquad \frac{r}{n}\frac{dn}{dr} < -1 ,$$

where $n$ is the refractive index at radius $r$. The Lorentz-Lorenz form of the Clausius-Mosotti relation is

$$\frac{1}{\rho}\frac{(n^2-1)}{(n^2+2)} = \text{constant} ,$$

where $\rho$ is density and $n^2$ is dielectric constant. Differentiating gives

$$\frac{dn}{d\rho} = \frac{(n^2-1)(n^2+2)}{6n} .$$

Hydrostatic equilibrium specifies

$$\frac{dP}{dr} = -\frac{\rho GM}{r^2} ,$$

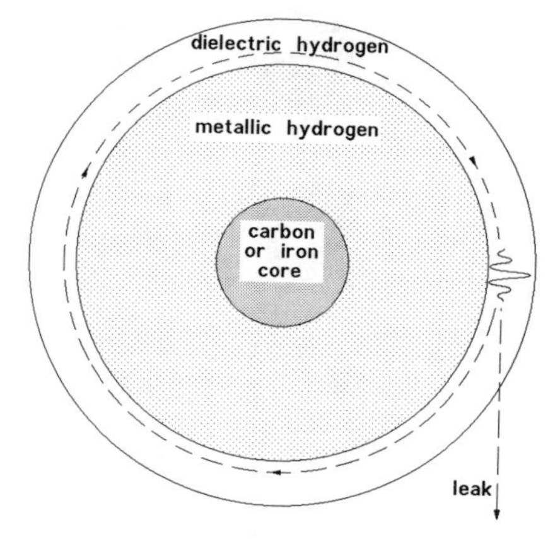


FIGURE 1. Ring cavity on degenerate dwarf. The leak will rotate with the pulse. (Not at all to scale.)

where $P$ is the pressure and $GM/r^2$ is the local gravity. Combining all of these formulae leads to

$$\frac{(n^2-1)(n^2+2)}{6n^2}\ \frac{GM}{r}\ \frac{d\rho}{dP} > 1 .$$

Various estimates for the phase-transition pressure of hydrogen differ considerably. Table I uses Ramsey's [1950] values that are adopted for our initial calculations. Accordingly, just above the metallic surface the radio refractive index should be about 1.6. More recent density estimates by Neece *et al.* [1971] would lead to a refractive index of 1.96, and the discrepancy may be indicative of the uncertainty. Inserting the figures into the final formula gives $GM/r > 2(10^{11})$ as the condition for inward spiraling rays. For a one solar mass dwarf, $GM/r$ is about $4(10^{17})$ and for the largest hydrogen dwarf it is about $10^{14}$, so the inequality is overwhelmingly fulfilled for all dwarfs. The fulfillment is so strong that waveguide-type propagation over the surface is a more apt description than distinct hops and bounds of the rays. The cavity acts like an inside-out whispering gallery.

TABLE 1. Estimated properties of compressed hydrogen

| Density (gm cm$^{-3}$) | Pressure ($10^{12}$ dyn cm$^{-2}$) | Refractive index |
|---|---|---|
| 0.07 | lab | 1.108 |
| 0.34 | 0.7 | |
| 0.35 | 0.8 | 1.62 |
| 0.77 | 0.8 | metallic |

FIGURE 2. Falling lasing dye droplets (negative display) [after Snow, Qian, & Chang].

The external appearance of this model can be expected to resemble that of the dye-droplet lasers shown in Figure 2. The droplets of dye solution drip from a capillary and are pumped by a flashlamp laser. The cavities are the whispering galleries given by total internal reflection just inside the surface. The closest resemblance should be for those droplets that are slightly oblate, defining an equator as the minimum-energy circumferential path. Note how the laser shines tangentially from that equator. The radiation within the cavity must become so strong as to perturb the closure of the cavity, permitting a small percentage to escape.

## Period Distribution

Figure 3 shows the theoretical mass-radius relations for degenerate dwarfs with the period distribution of pulsars superposed. There are two curves of the mass-radius relation that depend on the mean molecular weight of the core. Remembering that the core material is thoroughly ionized, the mean molecular weight in units of proton mass for hydrogen is $\frac{1}{2}$ whereas for heavy elements it is 2. Evolved dwarfs have exhausted their hydrogen by converting it to heavy elements, so they have the separate mass-radius relation. The peak of that curve is known as the *Chandrasekhar mass limit*. More will be said about the unevolved hydrogen dwarfs later in this discussion.

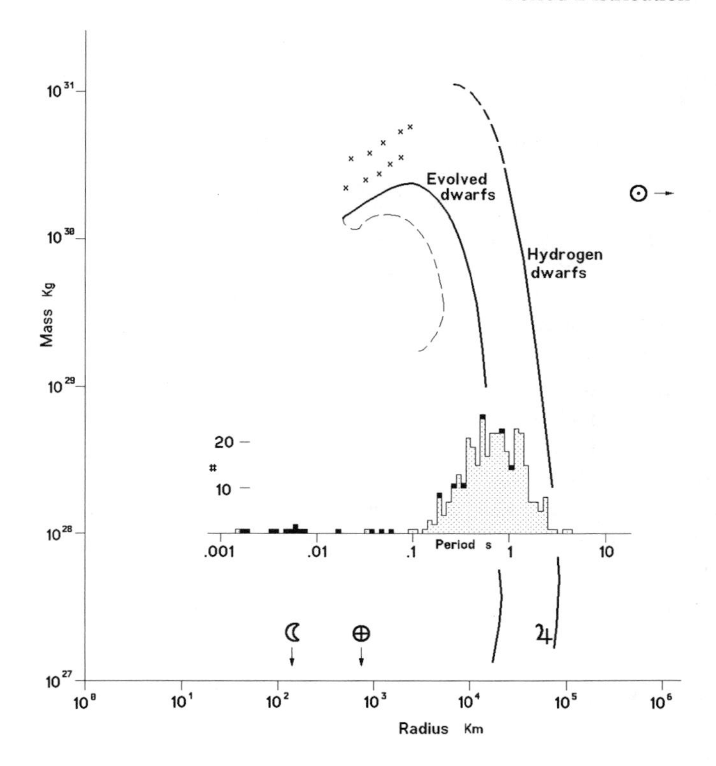


FIGURE 3. Period distribution of pulsars superposed on mass-radius relations for degenerate dwarfs. (x's are differentially rotating dwarfs: symbols indicate loci of earth, moon, jupiter, and sun: darkly shaded pulsars in the distribution are binary.)

Imagine starting with a fairly uniform assortment of masses at significantly larger radii. As these masses radiate away their thermal energy they will contract along horizontal paths until they reach the appropriate curve where they get stabily supported against further collapse by electron degeneracy pressure. The eventual number-radius distribution to be expected then will be proportional to the magnitude of the negative gradient of the mass-radius relation.

The characteristic periods of the cavities belonging to these dwarfs are simply their circumferences times the refractive index of dielectric hydrogen just before its transition to the metallic phase. The period distribution of pulsars is superposed on the mass-radius relations for dwarfs for comparison. The lateral alignment is dependent on the refractive index; it is set here at 1.96, but a change to 1.62 would shift the alignment by less than two bins, so there is no real flexibility to contrive agreements.

Most of the pulsars have periods in the range 0.1 to 3 seconds, and within that range the distribution at first appeared bimodal. Recently discovered pulsars, however, have tended to fill the gap. Those with periods less than 1 second had been classified by Huguenin *et al.* [1971] as having simple pulse shapes and match the period distribution expected from evolved dwarfs. Those with longer periods tend to have complex pulse shapes and can be attributed to unevolved hydrogen dwarfs.

Unevolved hydrogen dwarfs have been given the name brown dwarfs in contrast to the original white dwarf stars; both have degenerate interiors. Our major planet, Jupiter, fits the category of brown dwarfs. It has been suggested that brown dwarfs might contribute most of the "missing" mass of galaxies. They are small and not brightly self-luminous and would be exceedingly difficult to observe. None had been clearly identified outside the solar system, but very recently a pulsar with planetary companions of small mass has been discovered. Just as recently, gravitational lensing events that brighten stars in the Magellanic cloud have been detected and attributed to brown dwarfs. Those sought for the missing mass are distinctly less massive than those that might serve here as long-period pulsars. Why? The least massive star known is about one-tenth of a solar mass, and so astronomers presume that more massive configurations all will ignite nuclear reactions to become ordinary stars.

## Period-luminosity Cutoff

Consider, however, a progenitor that is radiating energy and contracting toward becoming a star. The virial theorem tells us that under gradual contraction gravitational energy remains in balance with twice the internal kinetic energy. A rotating configuration should not get as hot as a nonrotating one because rotational kinetic energy can replace thermal kinetic energy. The point is that a rotating configuration may not get hot enough for nuclear ignition. Although authorities have assured me that the effect is truly negligible, it should at least be there qualitatively and we do know that nuclear ignition is extremely sensitive to temperature. If it is not negligible, the consequences are interesting.

For one thing, we will be underestimating the number of masses above one-tenth of a solar mass because only the nonrotating ones will have ignited, and that will be especially so just above that mass. Thus the extrapolation of the number of masses to those below one-tenth of a solar mass may be grossly underestimated. We also know that stars of later than F spectral class are never rapid rotaters. These are relatively light as well as relatively cool stars along the main sequence. Conventional wisdom explains that observation as the consequence of stellar winds, the idea being that the winds would carry angular momentum away from such stars. Earlier (hotter and more massive) stars, without winds, would have no such way of slowing their rotations. On the other hand, it might just be, at least in part, that the rapid rotators never ignited to become full-fledged stars. A factor that would contribute to this effect is that it is difficult to transport angular momentum across a gradient of molecular weight. Thus, if chemical fractionation develops within the progenitor, the internal rotation will tend to be preserved. I am not familiar with the details of the phenomenon, however, and so can only mention it in passing.

Progenitors more massive than the sun will always end up igniting because their central pressures get so high that pyknonuclear reactions (pressure induced) can occur. This would be why the gap at 1 second in the period distribution of pulsars is not filled in by massive hydrogen dwarfs.

## Short Periods

Apart from the bimodal distribution peaks there is a whole string of fast pulsars with periods down to 1.6 milliseconds. These also must be accommodated, but they are to the left of the Chandrasekhar peak. Being to the left, they cannot have arrived there by gradual contraction along a horizontal path preserving mass, because they would have found stability on the right side of the peak. They must have gotten there via a downward path, implying mass loss during formation. Two of those, the Crab and Vela pulsars, are clearly supernova remnants and therefore easily qualify for rapid mass loss. We may wonder about the stability of these configurations because initial calculations suggested that they would be unstabile against vibration. In the early 1960s, however, Ostriker [1971] found that rotation stabilized them, and differential rotation stabilized them even more. Furthermore, he found that rotation and differential rotation somewhat increased the mass of the Chandrasekhar limit. The configurations shown by crosses in Figure 3 were calculated by Gribbin [1969] and found to be stable.

Most of the other fast pulsars are binaries. One route to binary status is stellar fission. A mass above the Chandrasekhar limit will contract to an unstable configuration and so will break in two from the growing vibration. However, for a two-body situation, gravity still will bring the pieces to fall back upon themselves. The resulting collision then can make a three-(or more)-body situation where one can depart, leaving the others in orbital stability. Another possibility is for magnetic forces to avert the collision. A popular scientific toy in the 1960s was a pendulum with a magnetic bob. Fixed magnets could be placed just under the nadir of the pendulum so as to deflect the motion to curious unpredictable paths. Such might well be the case for the fission of magnetic stars. If so, fission is a good way to lose one-half of the mass in a hurry, and so qualify for arrival on the left side of the dwarf stability peak.

Orbital dynamical considerations constrain the component masses of the binary pulsars to the approximate range 1 to 1.5 solar mass. Such masses are just as consistent with dwarfs as with neutron stars, and so those considerations do not at all serve to discriminate between the two. Even anticipated and observed effects of gravitational radiation leading to the orbital precession of PSR 1913+16 do not discriminate between the two. The popular explanation of pulsar recycling with accretion-driven spinup does seem rather contrived to me, however. That explanation says that we see them during the recycling process no longer as radio pulsars but as low-mass X-ray binaries, because we still do not see any radio pulsars speeding up. And we may still ask why we do not see any pulsars as binaries prior to the recycling process.

Furthermore, there is a quite recently discovered pulsar that has at least two companions that have planetary size masses. It is very difficult to expect that such a system can be the remnant of a supernova explosion simply because it should have blown apart. It is just as hard to imagine how it might have formed thereafter. On the other hand, a fission process might very well end up with several planetary mass components since that offers three-(or more)-body dynamics, rather than just magnetic repulsion, to escape falling back on itself.

A number of the pulsars have periods that are still less than 20 milliseconds, where the mass-radius relation appears to hook back toward a larger radius than 500 kilometers. A complaint about this cavity model is that there should be no pulsars at such short periods. Inasmuch as the theoretical curve ignores both rotation and magnetism, both of which carry a stabilizing influence, the complaint should be noted but not deemed to be decisive.

Although most of those short-period pulsars are binary, there are several that are single. Of these, all but one, the shortest, are found in globular clusters. The density of stars in globular clusters is sufficient enough that chance encounters which can easily separate a binary should be frequent.

The pulsar having the shortest period of 1.558 milliseconds would be too small to belong among the dwarfs. The cavity size would place it as a neutron star. Although the surface layers of a neutron star would be thinner than those of a dwarf because of the higher gravity, otherwise there should not be much difference, and so a cavity still is quite plausible. The gravity is actually so high that the observed periods will be lengthened by the gravitational redshift factor $(1+Z)$. Adopting the sizes published by Durgapal *et al.* [1983] based on the rather massive neutron stars favored by Brecher and Caporaso [1976] and correcting for the red-shift gives periods in the range 1.42 to 1.51 milliseconds. Curiously, the longer of those periods corresponds to the smaller model because of the red-shift. The few percent discrepancy of these periods with the actual 1.558 milliseconds easily could be accounted for by rotational oblateness. Rotating models would be somewhat more massive and slightly larger than those published, just as is found for dwarfs. The surprisingly close agreement ought not be taken too seriously, however, since there is considerable uncertainty in the mass-radius relationship for neutron stars [Baym and Pethick 1979].

## Slowdowns

Having established the existence of suitably sized cavities, it becomes of vital importance to account for the systematic slowdown of pulsars. Why should the cavities grow in size? Our dwarfs are presumably not very hot, nor are they absolutely cold. Their thermal radiation will result in gradual cooling leading to an even more gradual contraction. A first impression might suggest speedups rather than slowdowns. But think how the contraction takes place. As it cools down it becomes easier to compress the dielectric hydrogen to the metallic phase. The precipitation then will enlarge the metallic sphere. It is much like having a glass of water with sand stirred up in it. As the sand settles, the overall center of gravity lowers but the sand bottom rises.

This mechanism of precipitation is the same as that proposed by Smoluchowski [1967] to explain the infrared luminosity of Jupiter. Jupiter radiates about twice as much energy as it receives from the sun. The excess is far too much to be explained as either primordial or coming from radioactivity. Smoluchowski deduced that the gravitational energy released by precipitation causing a core radius increase of roughly 2 millimeters/year would suffice. That is remarkable dimensional stability for something that big. Furthermore, the argument can be turned around to say that any worse stability would require substantial transfers of energy. On the basis of those figures, pulsar radii are typically increasing by about 2 meters/year, and the Crab pulsar would be generating about one solar luminosity. That is quite insufficient as a source of radio pulse power.

Occasional flares or flare-like storms would provide the observed discrete speedups by cooking off some of the metal, and general temperature fluctuations might provide the small but real period irregularities [Horowitz *et al.* 1971].

## Energy and Pumping

Maser or laser action still requires a souce of energy and a pumping mechanism in addition to the cavities for operation. As was already mentioned, many of the dwarfs under consideration have to be rotating for structural stability, and, furthermore, the rotation is likely to be differential rotation. Although small, these dwarfs are about 1000 times larger than neutron stars. Since they both have roughly the same masses, the dwarfs can hold much more rotational kinetic energy even at lower rotational rates. The lower rates are sustainable by the dwarfs and hence are too low for being the pulsar clock, but they still easily suffice for the energy supply. Their energy supply is more than 100 times that attributed for rotating neutron stars, and so commensurately lower pulse generation efficiency can be tolerated.

As for the pumping, we have grown too accustomed to thinking of laser action exclusively in terms of atomic energy levels with population inversion. These supply narrow-band gain, but for pulse generation we will be more concerned with wide-band gain and ought not feel uncomfortable about thinking of electrical instability as a broad-band gain mechanism. The very definition of instability is that if you tweak it, the response is excessive, amplifying the stimulus.

Considering the physical circumstances, it should not be at all unreasonable to expect electrical instability. For one thing, there is a strong electric field in the dwarf, because it is the electron pressure supporting the material against the gravity that acts more on the protons. The thing that keeps the protons from falling and the electrons from bursting out is the electric field holding them together. Ordinarily we are accustomed to currents accompanying electric fields in conductors, but such need not be the case here because of the balance of gravity.

Then there is the internal magnetic field, which will be wound up into a tight spiral as it gets dragged along by the differential rotation of the metallic hydrogen. That in itself should be an unstable situation, since the dragging absorbs work into the magnetic field that in turn would like to dissipate that work by straightening out.

Then there is the problem of the Meissner effect. It is thought that metallic hydrogen is superconducting [Ashcroft 1968]. In that case, the magnetic field does not just get wound up, it should get completely expelled. Nevertheless, we know that Jupiter does have a magnetic field even though it also presumably has a core of metallic hydrogen. All in all, the propects for electrical instability are quite plausible. Thus the reflective surface of our cavity could plausibly have a reflection coefficient slightly greater than unity that serves as the broad-band stimulated emission implicit for maser action.

## Pulse Formation

Given the cavities and power mechanism, the next step is to describe the pulse formation that is characteristic of laser mode-locking and that gave rise to the inception of this scenario in the first place. The process, contrary to the rather forbiding sound of its name, turns out to be quite simple and natural. An easy way to understand the process comes from its radio version, Cutler's regenerative microwave pulse generator [1955]. Figure 4 shows its block diagram. At first let us ignore the nonlinear element. When the loop gain exceeds unity, a periodic waveform builds up. Those frequencies that have an integral number of waves in the loop delay are constructively reinforced for the buildup to become the harmonics of the periodic waveform whose period matches that delay. That is effectively how an ordinary laser works, except that ordinarily the gain is exceedingly narrow-band so that only one harmonic (mode) gets adequately amplified. For broader band gain several or many harmonics participate to give a complicated but still periodic waveform. We now can consider the nonlinear element, which is typically a saturable absorber that hinders the low amplitudes more than high amplitudes. Consequently, those weaker amplitudes decay relative to the stronger ones, and the peak amplitude of the waveform wins out to become a pulse. The phases of the modes then are said to be locked, and the mechanism is called *passive*

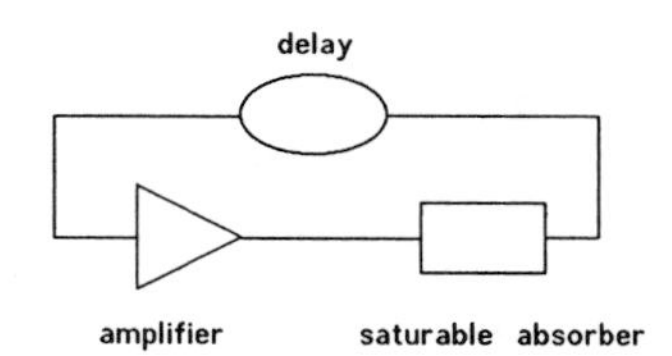


FIGURE 4. Cutler's regenerative microwave pulse generator.

*mode-locking*. Since all absorbers are ultimately saturable, passive mode-locking is almost inevitable for broad-band multimode lasers.

Another possibility occurs when the nonlinear element is incorporated directly into the amplifier as a saturable amplifier. Figure 5 typifies the operation, where a flashlamp acts as the saturable amplifier. When the flashlamp is triggered it exhausts its stored energy as a flash. The round-trip time of the light to the mirror allows for the energy recharge of the flashlamp, permitting it to be retriggered by the photocell at the return of the light pulse. This mechanism is called *self pulsing* in lasers. Again it is a common phenomenon in lasers.

Occasionally, very long lasers have been built by placing mirrors on neighboring hilltops [Linford *et al.* 1974]. They are so long as to have $10^9$ waves in their cavities, just as with our pulsar cavities. Quite unintentionally, all of these lasers turned out to be mode-locked, simply because it is a most natural phenomenon.

As for the waves escaping the cavity, when the gain exceeds unity the intensity in the cavity builds up exponentially with time until it is so strong that it perturbs the index gradient that closed the cavity in the first place. That perturbation then can allow tangential leakage at the pulse site and thereby present a rotating beacon. That some such leakage can actually take place is evident from the appearance seen in Figure 2.

The most obvious resemblence of pulsar pulses to laser pulses is in the duty cycle. For both, the pulse typically occupies about 10 to 15% of the period. Then there are the interpulses. These result from transverse modes propagating slightly skew to the principal longitudinal mode. The interpulses are not necessarily half-way between the main pulses. Figure 6 compares the pulse profiles of pulsars and lasers.

## Mode-hopping

PSR 0329 presents a particularly curious interpulse behavior [Lyne 1971]. It has both a leading and a trailing interpulse symmetrically disposed around the main pulse. Ordinarily, the trailing interpulse is twice as strong as the leading one. From time to time, however, it abruptly changes so that the leader becomes twice as strong

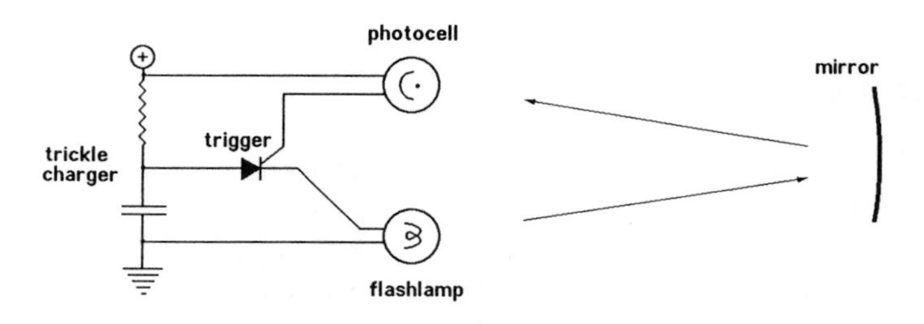


FIGURE 5. Self-pulsing mechanism.

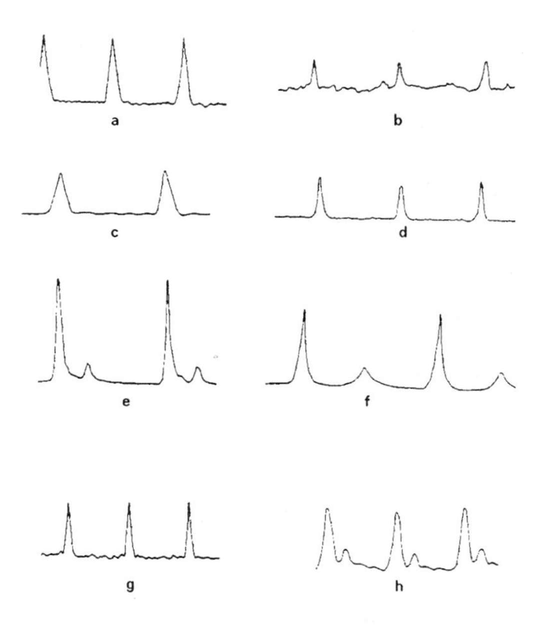


FIGURE 6. Comparative pulse profiles for lasers and pulsars: *A:* He-Ne laser, *B:* PSR 1133+16, *C:* CO laser, *D:* PSR 0950+08, *E:* He-Ne laser, *F:* PSR 0531+21, *G:* He-Ne laser, *H:* Nd:YAG laser.

as the trailer. The changes are very abrupt, taking place within about one period, and the duration of the abnormal profile lasts for tens or hundreds of pulses with no regularity. That behavior is readily interpreted as instability and competition of complementary transverse modes for a laser, whereas it is difficult to reconcile with rotating neutron stars.

Implicit in the specification of transverse modes is a principal optical path. The oblateness and the equator from the rotation of the dwarf give that path. Figure 2 showed that the radiation comes off such an equator tangentially. For pulsars, gyroscopy and Fresnel dragging would remove the ambiguity of direction around the equator.

## Polarization

The polarization behavior of the pulses is also interesting. Typically, the polarization sweeps through about 90° during the course of a pulse. Only a very weak birefringence, perhaps due to a magnetic field, need be assumed to explain this. The slight difference in circumferential travel times would pull the pulse forward for one polarization while dragging it back for the other. The nonlinear element responsible for the mode-locking should be a function of the composite amplitude and so holds the two together at a mean period; but still the composite pulse will start off in one polarization and end up in the other. Furthermore, for the Crab pulsar, the interpulse mimics the polarization sweep of the main pulse. That is quite easy to appreciate since both pulses traverse essentially the same birefringence. To the best of my knowledge, this topic about polarization has not been investigated in lasers, but it is certainly plausible.

If there is more birefringence present, then the two polarizations can beat within the pulse. Those beats would appear as subpulses, which can seem to march with successive pulses and have the same polarization sweeps on each of the subpulses, as was discussed for the main pulses. Precisely that behavior is conspicuosly exemplified by the observations of PSR 0809+74 as shown in Figure 7. A significant number of other pulsars also display marching subpulses, although their polarizations have not been measured. The speed of laser pulses makes it difficult to resolve subpulses. In spite of the difficulty, the results shown in Figure 8 clearly show subpulses. Whether or not the lasers show marching behavior or polarization sweeping remains unknown.

## Visible and X-ray Pulses

A couple of pulsars, the Crab and Vela, generate, in addition, visible and X-ray pulses that are synchronized with their radio pulses. Surely the cavity does not have sufficient precision to function at such short wavelengths. The spectrum is

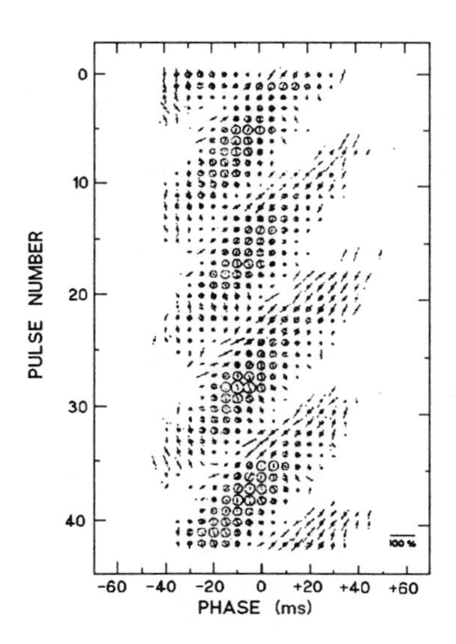


FIGURE 7. Polarized phase-time diagram for the drifting subpulses from PSR 0809+74 (after Taylor et al. 1971).

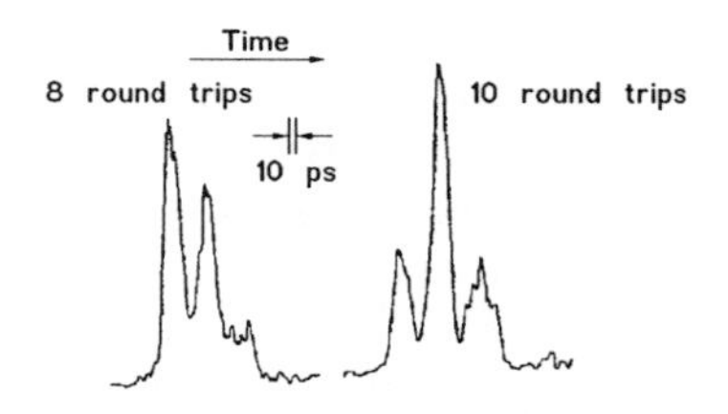


FIGURE 8. Two next-successive pulses from a passively mode-locked dye laser. [Arthur, Bradley, & Roddie 1973]

not continuous, however; they do not have microwave pulses. Those two pulsars, unlike other pulsars, are clearly identified as supernova remnants. Certainly the Crab, but probably the Vela as well, reside in a cloud of high-energy electrons. The radio pulses are extremely intense in the immediate vicinity of the pulsars and could interact with those electrons to give stimulated inverse Compton scattering [Pantell *et al.* 1968] or stimulated bremsstrahlung [Madey 1971]. The resultant visible and X-ray pulses then would be coherent and synchronous with the radio pulse.

For the Crab pulsar, its cloud of high-energy electrons would be expected to dissipate and so must be continually refurnished. It is not clear to me just how that is accomplished for this model, or for any other model.

## Jovian Decametric Emission

Jovian decametric is certainly enigmatic and may or may not be related to this pulsar model. On the may not side, its radio frequencies are lower than those of pulsars by about a factor of ten [Carr *et al.* 1964]. That may, however, just be the result of solar-induced weather that disturbs the quality of the cavity and renders it unfit for the shorter wavelengths. It should be said that Hirshfield and Bekefi [1963] did propose maser-like emission from high-energy electrons , although the most popular explanation is that of Goldreich and Lynden-Bell [1969].

The depth at which the transition to metallic hydrogen occurs in Jupiter is fairly uncertain, but a rough guess suggests a characteristic period of about 1.5 to 2.5 seconds. If so, we might expect intensity fluctuations of the decametric radiation at that period. Although there have been many measurements [e.g., Douglas and Smith 1967] made of the decametric radiation, none that I have seen are on a suitable scale for looking. The published observations are unfortunately either too quick or too slow.

If there is trapped radiation leaking out of the cavity, it would appear to come from a crescent on the limb of Jupiter, much like that which appears in Figure 2. East-west interferometric observations [Dulk *et al.* 1967] indicate that the apparent source size is less than 0.1 arcminute. A couple of north-south observations [Brown *et al.* 1968; Slee and Higgins 1966] indicated sizes ranging from 6 to 15 arcminutes, consistent with the crescent appearance, but these latter observations have not been confirmed [Lynch *et al.* 1972].

One very puzzling phenomenon of the Jovian radiation is its intensity correlation with elongations of Io [Bigg 1964]. The tidal distortion of Jupiter by Io could serve as an elegant valve on the cavity that would release the radiation primarily at Io elongations. An observational clue may be found in the frequency drifts of the radio bursts associated with the elongations. The valve would leak the radiation just as a dispersive prism, so as to give opposite frequency drifts for east and west elongations, just as observed.

All of the decametric observations are extremely complicated, however, with L bursts and S bursts and associations with locations A,B,C on the planetary surface [Carr and Gulkis 1969]. Magnetospheric physics is never simple, and it is complicated here by the interaction of the Jovian magnetic field lines with Io, so that there very well may be no relation with pulsar radiation.

## Summary

In summary, it is really the two main features that primarily recommend this mode-locked maser interpretation of pulsars. First, we are apprised of the the existence of natural cavities at the surface of ordinary degenerate dwarfs, and that these cavities have characteristic periods that specifically match the period distribution of pulsars. Rotating neutron stars, on the other hand, might have any periods from milliseconds on up and so are merely consistent with the periods of pulsars. It seems to me that the better specificity favors the cavity model over rotation, especially since the latter was deduced by the process of elimination from an incomplete set.

Second, we see an authentic pulse-generating mechanism that not only assuredly makes pulses, but it also makes a whole rococo morphology of pulses which is indistinguishable from that of pulsar pulses. It therefore seems to me most likely that their mechanisms are one and the same. In contrast, the pulse emission mechanism, let alone the pulse morphology, for rotating neutron stars remains an embarrassing mystery that still defies any widely convincing explanation after all of these years.

Since there has not been any compelling evidence that would rule out the maser model, it seems to me that at the very least it ought to deserve an open place on the table for consideration even though it does disagree with deeply entrenched dogma.

## Bibliography

E. G. Arthurs, D. J. Bradley and A. G. Roddie, 1973 *Appl. Phys. Lett.* 23: 88

N. W. Ashcroft, 1968 *Phys. Rev. Letters* 21: 1748

G. Baym and C. Pethick, 1979 *Ann. Rev. Astron. and Astrophys.* 17: 415

E. K. Bigg, 1964 *Nature* 203: 1008

P. V. Bliokh, A. P. Nicholaenko and Yu. F. Fillippov, 1980 *Schumann Resonances in the Earth-Ionosphere Cavity,* Peter Peregrinus Ltd.

K. Brecher and G. Caporaso, 1976 *Nature* 259: 377

G. W. Brown, T. D. Carr and W. F. Block, 1968 *Astrophys. Letters* 1: 89

T. D. Carr, G. W. Brown and A. G. Smith, 1964 *Astrophys. J.* 140: 778

T. D. Carr and S. Gulkis, 1969 *Ann. Rev. Astron. Astrophys.* 7: 577

M. H. Crowell, 1965 *IEEE J. Quant. Elec.* QE-1: 12

C. C. Cutler, 1955 *Proc. IRE* 43:140

J. N. Douglas and H. J. Smith, 1967 *Astrophys. J.* 148: 885

M. A. Duguay, S. L. Shapiro and P. M. Rentzepio, 1967 *Phys. Rev. Lett.* 19: 1014

G. Dulk, B. Rayhrer and R. Lawrence, 1967 *Astrophys. J.* 150: L117

M. C. Durgapal, A. K. Pande and P. S. Rawat, 1983 *Astrophys. and Space Sci.* 90: 117

A. A. Fife and S. Gygax, 1972 *Appl. Phys. Lett.* 20: 152

A. F. Gibson and M. F. Kimmitt, 1972 *Laser Focus (August)* 26

V. L. Ginzburg, 1971 *Physics* 55: 207

J. R. Gribben, 1969 *Astrophys. Lett.* 4: 77

P. Goldreich and D. Lynden-Bell, 1969 *Astrophys. J.* 156: 59

B. K. Harrison, K. S. Thorne, M. Wakano and J. A. Wheeler, 1965 *Gravitational Theory and Gravitational Collapse,* U. Chicago Press

J. L. Hirshfield and G. Bekefi, 1963 *Nature* 198: 20

P. Horowitz *et al.* 1971 *Astrophys. J.* 166: L91

G. R. Huguenin, R. N. Manchester and J. H. Taylor, 1971 *Astrophys. J.* 169: 97

L. Leopold, W. D. Gregory and J. Bostok, 1969 *Can. J. Phys* 47: 1167

G. J. Linford and L. W. Hill, 1974 *Appl. Opt.* 13: 1387

G. J. Linford *et al.* 1974 *Appl. Opt.* 13: 379

M. A. Lynch *et al.* 1972 *Astrophys. Lett.* 10: 153

A. G. Lyne, 1971 *Mon. Not. Roy. Astron. Soc.* 153: 27

A. G. Lyne and F. G. Smith, 1968 *Nature* 218: 124

J. M. J. Madey, 1971 *J. Appl. Phys.* 42: 1906

R. N. Manchester and J. H. Taylor, 1977 *Pulsars* (W. H. Freeman and Co.)

L. Mertz, 1974 *Astrophys. and Space Sci.* 30: 43

F. R. Nash, 1967 *IEEE J. Quant. Elec.* QE-3: 189

G. A. Neece, F. J. Rogers and W. G. Hoover, 1971 *J. Comp. Phys.* 7: 621

G. H. C. New, 1972a *Alta Frequenza* 41: 711

G. H. C. New, 1972b *Optics Comm.* 6: 188

J. P. Ostriker, 1971 *Ann. Rev. Astron. Astrophys.* 20: 353

R. H. Pantell, G. Soncini and H. E. Putoff, 1968 *IEEE J. Quant. Elec.* QE- 4: 905

C. Papaliolios, N. P. Carleton and P. Horowitz, 1970 *Nature* 228: 445

J. D. H. Pilkington *et al.* 1968 *Nature* 218: 126

A. J. R. Prentice and D. TerHaar, 1969 *Mon. Not. Roy. Astron. Soc.* 146:423

S. Qian, J. B. Snow, H. Tzeng and R. K. Chang, 1986 *Science* 231: 486

W. H. Ramsey, 1950 *Mon. Not. Roy. Astron. Soc* 110: 444

H. Risken and K. Nummedal, 1968 *J. Appl. Phys.* 39: 4662

M. Ruderman, 1972 *Ann. Rev. Astron. Astrophys.* 10: 427

W. Sieber and R. Wielebinski, 1981 *Pulsars* IAU Symp. 95 (W. Reidel Publ. Co.)

O. B. Slee and C. S. Higgins, 1966 *Australian J. Phys.* 19: 167

P. W. Smith, 1968 *Appl. Phys. Letters* 13: 235

P. W. Smith, 1970 *Proc. IEEE* 58: 1342

R. Smoluchowski, 1967 *Nature* 215: 691

J. B. Snow, S. Qian and R. K. Chang, 1986(May) *Optics News* 12: 5

D. A. Soderman and W. H. Pinkston, 1972 *Appl. Opt.* 11: 2162

S. Tanaka and K. Takayama, 1967 *J. Phys. Soc. Japan* 22: 300 and 311

J. H. Taylor *et al.* 1971 *Astrophys. Letters* 9: 205

HP. Weber and R. Ulrich, 1971 *Appl. Phys. Letters* 19: 38

# 7

# Quasars

## Introduction

The discovery of quasars in the early 1960s marked a turning point in astronomy and opened for us a vastly more immense vista on the universe. The tale of that discovery is completely fascinating, being best told by those closely involved. The main detail necessary here is that quasars appear as point-like optical sources, usually identified with radio sources, and having optical emission lines at large redshifts. The redshifts turned out to be much larger than anything seen previously, and if attributed to the Hubble expansion of the universe implied enormous distances and enormous energetics. Furthermore, it soon was discovered that many of them displayed variable brightness on time scales of weeks or less. That has been especially awkward for astronomers to understand since a source is not supposed to be able to turn on or off in a shorter time scale than the light travel time across the source. That should mean that sources are very small for their power, and yet there appear some forbidden emission lines only to be expected from low-density material.

Somewhat more recently another puzzle has developed. Quasars are found too frequently in proximity to galaxies. Although most astronomers insist that it is just happenstance, the evidence is conspicuous and the topic is highly controversial [Arp 1987].

Further puzzles have arisen as the new capabilities for imaging at radio wavelengths have revealed jets, frequently double, not always straight, and often with nebulous lobes, connected with the quasars. At even higher angular resolution in the milliarcsecond range the quasars frequently appear to be double. Repeating the observations a few years later, the separation of the double has noticeably increased. In all cases, the separation always increases; no decreases are seen. Assuming the distance to the quasar based on the redshift, we casually estimate the velocity across the line of sight to be typically ten times the speed of light. Surely that is an illusion. Circumstances for such an illusion are that a component of the double is going away from the other in a direction almost directly toward us at a significant fraction of the speed of light. While this illusion is easily accepted as the explanation, it does leave the puzzle as to why so many things are aimed almost directly at us.

## Superluminal Illusion

Let us see how these pieces of the quasar puzzle can fall into place from an explanation of this superluminal illusion. The nature of the superluminal illusion, although not widely known, is nevertheless well known as basically the same as that of light echoes. Our awareness of light echoes comes from Nova Perseii 1901. In a matter of seven months a growing nebulosity appeared. At first it was thought that the nova was rather close by, for otherwise the expansion rate would appear to exceed the velocity of light. Only much later did the light echo get explained.

The nova emits a flash of light that expands radially. At first we might naively think that the locus of what we would see at any instant would be a spherical shell around the nova. On second thought, however, we realize that what we see lies on an ellipse with the nova at one focus and ourselves at the other. At our large distance the ellipse closely approximates a parabola whose polar equation is $r = p/(1 - \cos\theta)$, where $p$ is the semilatus rectum and $\theta$ is the angle to the line of sight. Figure 1 shows the geometry. The naive spherical interpretation would have it that $p$ should be the apparent maximum radius. The more cor-

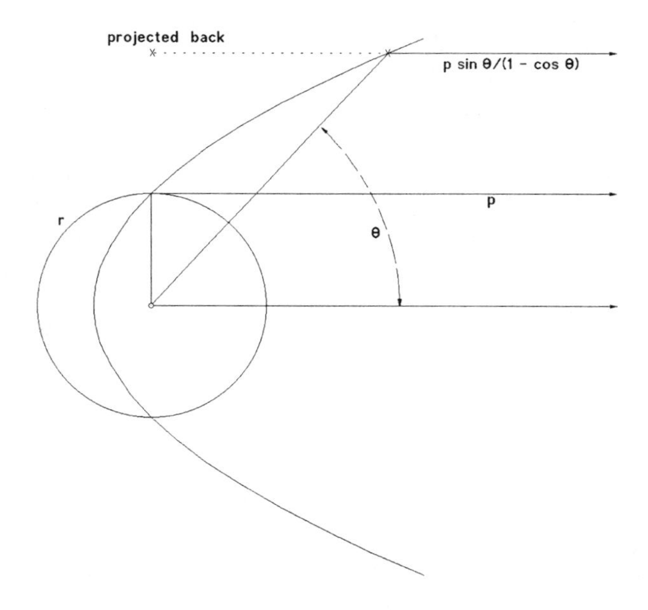


FIGURE 1. Geometry for light echoes and the superluminal illusion.

rect interpretation gives $r \sin\theta$, which imparts the factor $\sin\theta/(1 - \cos\theta)$ that can be significantly greater than 1 to give the illusion of superluminal expansion.

If the actual expansion velocity is less than that of light, then $r$ is diminished by $\beta = v/c$, but the overall result still can appear to be superluminal. Thus the standard explanation for the apparently superluminal observations of quasars is that they eject matter almost directly toward us at some fraction of the speed of light. I will now argue that it need not be physical matter that is ejected, but simply a gravitational image. This argument was inspired by a view of the sun reflected off the Pacific Ocean. The photograph in Figure 2, although off of a lake, shows the salient features.

## Gravitational Lensing

The first thing to notice is that the image is fragmented into glints. Unlike the speckles of speckle interferometry, the glints are achromatic. The glints are found to a certain degree off the optical axis and then there are no more. That is simply because there is a limit to the tilt angles of the water surface. The apparent size of the individual glints is noticeably less than the half-degree apparent size of the sun.

FIGURE 2. Solar reflection broken into glints (courtesy G. Davidson).

An important attribute is that the glints are localized close to the water surface. It would be a deceptive mistake to project them back to the distance of the sun. Finally, if we were to observe them as a function of time, they turn on and off very quickly, certainly much faster than the 2.3-second light travel time across the radius of the sun.

While these attributes are indicative, it is very easy to create an even more realistic simulation of the aspects of gravitational lensing. Contrary to the behavior of ordinary lenses, gravitational lenses bend the light more near the center of the lens than at the periphery. Their effective geometry is much like the base of a wine glass broken off at the stem. Perhaps the shape should be called a "cuspoidal axilens." Since the discovery of the first convincing example of a gravitational lens several observatories have made plexiglass models based on detailed calculations, but even the inexact contours of the wine glass base well illustrate the salient aspects of the lensing.

The first discovered example was QSO 0957+561 [Walsh *et al.* 1979], and the convincing evidence was that the A and B components showed identical spectra. Dynamically related components would at least have been expected to show distinctive Doppler shifts and different line strength ratios. The discovery gave credibility to the earlier analysis by Barnothy and Barnothy [1968] indicating that occurences of lensing by galaxies should not be rare. Figure 3 illustrates the scenario envisaged here.

The existence of one case stirred interest, and the prospects for smaller multiple cases came to be considered because of the hierarchy of the universe. There are many more smaller objects than there are larger objects. A particular paper that attracted my attention [Kayser *et al.* 1986] showed lovely caustic illumination patterns projected through an assemblage of gravitational lenses. Those patterns, calculated with numerical simulation, had important ramifications that I will return to shortly.

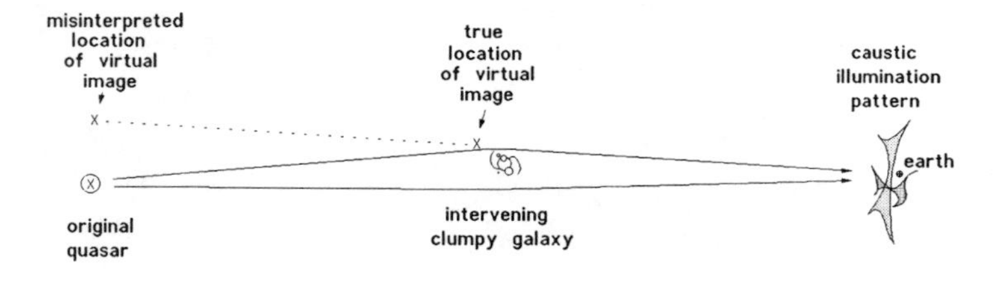


FIGURE 3. Scenario of gravitational lensing.

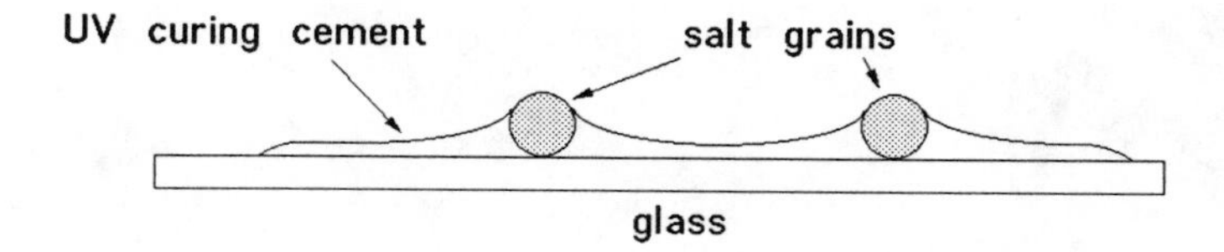


FIGURE 4. Experimental lensing screen.

## Experimental Simulation

For the moment, here is a recipe for simulating an ensemble of gravitational lenslets. Sprinkle a few grains of table salt onto a flat piece of glass. Add a thin layer of UV curing cement. The surface tension of the cement conforms the surface over each grain like a canopy having the shape of a cuspoidal axilens. When it looks satisfactory, the cement can be permanently cured with ultraviolet light. Figure 4 shows qualitatively that a section has the desired form. To arrive at a more rigorous evaluation, one might solve the detailed capillary shape by integrating a differential equation.

The behavior of the screen is most vividly demonstrated by shining a laser pointer through it as in Figure 5, typically projecting lovely caustic illumination patterns as shown in Figure 6. This behavior does not require monochromatic illumination; any bright fairly collimated light source can be used.

The caustic patterns projected through the screen are just the same as those calculated by Kayser *et al.* [1986]. The example in Figure 5 showing such agreement between experiment and their calculation is reassuring. As they already noted, the character of the illumination pattern does change somewhat, but not really very much, with distance.

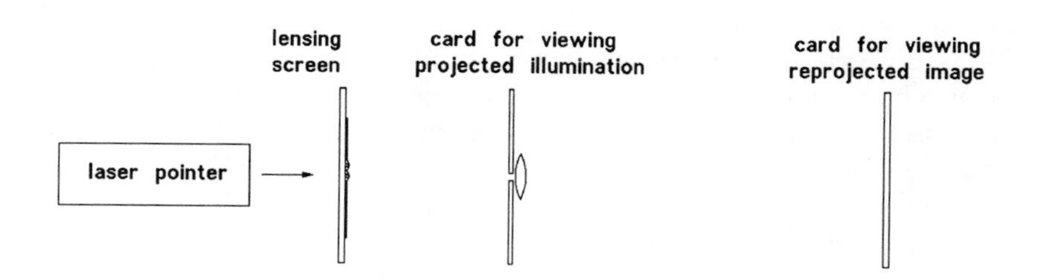


FIGURE 5. Projecting the illumination patterns.

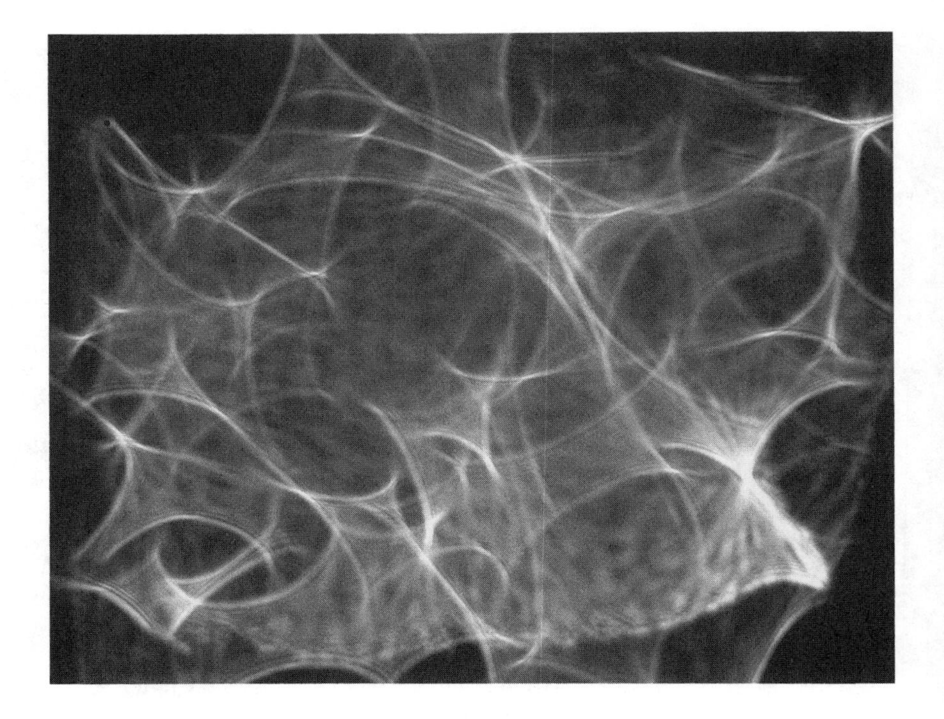

FIGURE 6. Caustic illumination pattern.

## Brightness Fluctuations

Kayser *et al.* also noted that transverse motions of any of the components would lead the earth across a slice of the projected illumination and consequently bring about brightness fluctuations. Inasmuch as the caustics have sharp edges, the fluctuations could appear to be quite rapid and not necessarily related to the light travel time across the original source. Furthermore, since the pattern would be convolved with the shape of the original source, the structure of the fluctuations might be indicative of projections of that shape. The action would be much the same as that for occultations of sources.

This origin of source fluctuations accords well with the observed temporal distribution function of quasar brightness fluctuations. For the most part, the sources are relatively dim and there are random occasional outbursts, just as a cut across Figure 6 would show. Furthermore, there is no apparent arrow of time for quasar fluctuations. Unlike for physical outbursts such as novae or gamma-ray bursts, which have sharp rises and slower declines, the time scale for quasar fluctuations can be reversed without noticeable change. Since we have no way of knowing in which direction we are slicing through Figure 6, there can be no such arrow of time distinctions. A slightly more formal way of looking at it is that the slope distribution function of physical outbursts shows a skew, while that for quasar fluctuations does not.

## Distance Scale Corruption

At a more primitive level, it is important to note that the presence of gravitational lensing adulterates the inverse square law that is so often used to deduce powers from distances. Since it would badly corrupt any brightness–redshift relation for quasars, it may be the reason for the lack of such a relationship. Much the same may be said for statistical number relationships, whether as functions of brightness or redshift. The existence of a steep gradient in the underlying statistics and a detection threshold upset the affair. There are more faint or high–redshift objects to be pushed up over the detection threshold than there are bright or low–redshift objects to be pushed under the threshold. The apparent pileup of objects just over the detection threshold upsets the slope that might be expected for a homogeneous Euclidian universe. On the other hand, we may note that the lensing merely rearranges the light, so it should not influence our perception of the 3°K cosmic background.

Looking back in the other direction, toward the screen, to see what we would see in the sky, we use a pinhole camera or a small telescope with a pinhole objective to give unlimited depth of focus. A typical result is shown in Figure 7, and again it is reassuring that this agrees with many of the actually observed candidates, especially with many of the radio images. However, it does not even attempt to agree with those spectacular radio sources having opposing jets that culminate in wispy nebular clouds. Those will have to seek explanation elsewhere.

## Foreground Localization

As Figure 7 shows, we do get little arcs and clumps. These should be more realistic than the glints of Figure 2 because the character of the screen better approximates that of gravitational lensing. It is an awful mistake to project the location of these clumps or glints back to the distance of the lamp. A careful look shows that each image clump is associated with a grain of salt on the screen, and that the clump is localized in the close vicinity of that grain. This localization is important because with the cosmological expansion of the universe the image clumps then are going away from the original source in a direction almost directly towards us at a modest fraction of the speed of light. It makes no difference for the superluminal illusion that the clumps and source are both going away from us.

This gravitational lensing hypothesis satisfies the observational constraints. In all cases of the illusion the quasars appear to be separating superluminally. As with meteor showers, in no instance do the meteors approach their radiant, and the geometry of the two situations is just the same. The quasar components separate out to a fraction of an arcsecond where the fainter one then fades away and vanishes.

FIGURE 7. Apparent source structure.

The geometry is quite automatic, and so it is not too surprising for the illusion to be so common. With the popular hypothesis of energetic mass ejection it is statistically hard to expect that in so many cases the ejection should be directed almost directly toward us. Why should the separations all be so fast? If we try to invoke a coherent beaming of the emission so that we only see those ejections toward us, then the universe must be even more energetic than our wildest imagination. All of the quasars would have to be ejecting matter in many directions all at once, and that is hard for me to believe.

Furthermore, there is no hint of relatively blue-shifted emission lines to correlate with the superluminal appearance. Yes there are the forest Lyman alpha absorption lines that are less redshifted than the emission, but these are a relatively passive consequence of intervening absorption clouds. I would expect at least some spectral emission associated with energetic mass ejection. On the other hand, if the component is merely a gravitational image rather than substantive matter, there would be no spectral shift to be expected.

With regard to the more widely spaced components that are resolvable in the optical regime, such as the original discovery QSO 0957+561, there has been the anticipation that a measured delay between the brightness fluctuations of the components would tell us about the distance and mass of the lensing galaxy. That is all very well if the fluctuations are intrinsic to the initial source; but if they happen to result from microlens structures of the lensing galaxy, that deduction would become faulty.

As others have suggested already, the observational effects of gravitational lensing should be used to assess the distribution of dark matter in the universe. In that sense we should avoid using any preconceptions about that distribution to dispute the phenomenon while that distribution still remains a puzzle. The observational effects will include the amplitude distribution of quasar variability; the statistics and sizes of arclets; and the statistics, relative component intensities, and maximum elongation of superluminal quasars. Hopefully a consistent picture might emerge from all that.

## Gravitational Amplification

But is refraction all there is to the effects of gravity upon light? There has been a longstanding controversy about whether the momentum of an electromagnetic field is given by $\mathbf{D}\times\mathbf{B}$ as favored by Minkowski or $\mathbf{E}\times\mathbf{H}$ as favored by Abraham. Although the problem has been around since 1914 and has been addressed by brilliant minds, it remains without a convincing resolution. For the most part, it should not make any discernable difference, but the reason that it might be important for cosmology stems from some sentences by Møller [1952]. He noted that if we accept Abraham's expression, then for light going through a moving dielectric "we have an exchange of energy between the electromagnetic and the mechanical systems, i.e., a local absorption and re-emission of light energy by the body. This clearly shows that Minkowski's separation of the total energy-momentum tensor into an electromagnetic part and a mechanical part is more natural than Abraham's, a transparent body being in Minkowski's theory a system that does not exchange energy even locally with the electromagnetic field." The existence of lasers was unknown at that time, but the re-emission in this exchange of energy must amount to a negative absorption or stimulated emission.

The momentum of a photon inside a dielectric would be $n$ (refractive index) times that outside according to Minkowski, but it would be $1/n$ times that outside according to Abraham. Since the forces exerted on the dielectric result from changes of momentum, their discrepancy even leads to opposing directions for the two theories. Ordinarily that is not important, since the immergent and emmergent forces cancel, but such is not necessarily the case for moving dielectrics.

The issue has become more muddled since Peierls [1976, 1991] concluded that the theories applied in combination, so that the sum of the two applied with extra confusion from dispersion. His conclusion also has been endorsed by Kastler [1974]. Experiments tend to confirm Minkowski, but not overwhelmingly because the experiments are both difficult to do and difficult to interpret [Jones 1988]. Something that puzzles me is that the Fresnel coefficient of drag is never cited as a criterion for distinguishing between the Minkowski and Abraham expressions. On the very same page as the earlier quotation from Møller, he states that Abraham's expression gives a result "corresponding to a dragging coefficient of twice the value given by Fresnel's formula." The drag has been measured by Michelson and by Zeeman, although it is again a difficult experiment and the potential errors are probably greater than admitted. A much more recent measurement [Sanders and Ezekiel, 1988] agrees with the Fresnel formula without mention of any possible factor of two. Mysterious factors of two do occur unexpectedly in relativity theory, and deliberate care must be exercised to avoid overlooking them. "The one thing harder than getting the sign correct in a physical calculation is getting the factors of 2 correct." [Strandberg 1986]. The ultraprecison interferometric techniques of Chapter 3 would be ideally suited for confirmatory experiments.

In brief, Abraham's theory leads to a different expression for the Fresnel coefficient of drag that also includes an imaginary component that would lead to an exchange of energy between electromagnetic radiation and moving dielectrics. The cosmological scenario is the same as for gravitational lensing, with the gravitational dielectrics between us and the very distant quasars. The Hubble expansion of the universe provides the motion that then might intensify the light from those distant quasars, and the energy for that intensification would be deducted from the enormous kinetic energy of that expansion. The enigma of quasars changes from anomalous red shifts to anomalous magnitudes.

As with gravitational lensing, the results would help answer several puzzles. First, there is Arp's observation that quasars tend to appear in the close neighborhood of galaxies. Quasars that otherwise would be too faint to see would get intensified by the gravitational fields of bodies expected in the neighborhood of galaxies. The main difference as compared to lensing would be a net intensification rather than only a redistribution. Wright [1982] pointed out to me that the net effect should then appear as a strong mottling of the 3°K black-body background. Such mottling is not observed, especially by the very sensitive Cosmic Background Explorer (COBE) satellite.

Despite this preponderance of evidence making Abraham's theory improbable, lingering uncertainties persist. Along with steady-state cosmology it conveys an obstinate aesthetic charm to confront the observations. Is it at all possible that the 3°K background is being misinterpreted, or that corrections used to compensate for known emissions from galaxies automatically erase the mottling? Whatever the case, it remains difficult to comprehend the extraordinary uniformity of the 3°K background in contrast with the very existence of galaxies, let alone clusters and superclusters.

One of the difficult problems in cosmology is how galaxies might have formed as condensations within the lifetime of the universe. While gravitation does attract objects together, unless there is some dissipative mechanism, they just as quickly fly apart. If Abraham's force were correct, then the interaction might transfer the kinetic energy into the radiation field and thereby provide the dissipation. While it is hard to envisage much efficacy to the process in our own environment, the extremes of radiation and motion in the early history of the universe or even near the cores of galaxies would accentuate the process.

# Bibliography


H. Arp, 1987 *Quasars, Redshifts and Controversies,* Interstellar Media

J. Barnothy and M. F. Barnothy, 1968 "Galaxies as gravitational lenses" *Science* 162: 348–352

R. V. Jones, 1988 *Instruments and Experiences,* Wiley

A. Kastler, 1974 *C. R. Acad. Sci. Paris B* 278: 1013–1015

R. Kayser, S. Refsdal and R. Stabell, 1986 *Astr. Astrophys.* 166: 36–52

C. Møller, 1952 *Theory of Relativity,* Oxford U. Press

R. Peierls, 1976 Proc. Roy. Soc. London A 347: 475
———1991 *More Surprises in Theoretical Physics,* Princeton U. Press

M. J. Rees, 1966 *Nature* 211: 468–470

G. A. Sanders and S. Ezekiel, 1988 "Measurement of Fresnel drag in moving media using a ring-resonator technique" *J. Opt. Soc. Am. B* 5: 674–678

M. W. P. Strandberg, 1986 "Special relativity completed: The source of some 2s in the magnitude of physical phenomena" *Am. J. Phys.* 54: 321–331

D. Walsh, R. F. Carswell and R. J. Weymann, 1979 *Nature* 279: 381–384

E. L. Wright, 1982 [private communication]

# Appendix 1

Appendix 1 The programs TELES, RPLAN, and MPLAN are Fortran code programs for the geometric design of aspheric reflecting systems. The READ(,) and WRITE(,) instructions may need to be adapted to the user's hardware. On mine, unit 9 refers to the console and unit 6 refers to the disk. TELES calculates the $x$, $y$ coordinates of a secondary reflector to correct the spherical aberration of a spherical primary. The origin of the coordinate system is the center of curvature of the primary.

Inputs are

R = radius of curvature of primary,
HM = maximum height of entrance ray on primary,
F = abscissa ($x$) of final focus,
PL = optical path from ordinate axis to focus.

Outputs are

entrance height, and $x$, $y$ coordinates of secondary.

Example inputs are

5.5, 3., 6.5, 12.3
5.5, 3., 2.75, 8.75
110., 60., 20., 207.

RPLAN calculates the $x$, $y$ coordinates of the secondary and tertiary mirrors for aplanatic telescopes (no spherical aberration and no offence against the Abbé sine condition) having a spherical primary. The coordinate origin is the center of curvature of the primary.

Inputs are

R = radius of curvature of primary,
HM = maximum entrance ray height (negative # signifies reverse direction at final focus),
SM = maximum sine of output ray,
FD = $x$ coordinate of final focus,
PL = path length from $y$-axis to final focus,
XD = approximate $x$-intercept of tertiary mirror.

Outputs are

entrance height, coordinates of secondary and tertiary.

Example inputs are

5.5, 3., .12, 4., 12.3, 1.385
5.5, 3., −.15, 4., 13.35, 2.5
5.5, −3., −.18, 0., 11.38, 1.6
6., −3., .3, 2.75, 9.53, 2.8.

MPLAN calculates the coordinates of a two-mirror aplanatic system, such as a microscope objective. Coordinate origin is the source focus.

Inputs are

AN = maximum sine of source ray,
G = magnification,
F = $x$ coordinate of output focus,
P = path length to output focus,
XD = approximate $x$ coordinate of secondary mirror.

Outputs are

sine of source ray, coordinates of two mirrors.

Example inputs are

.9, −10., 6., 7., 2.
.9, 8., 6., 6.2, 1.
.9, 8., 6., 6.2, 1.4

```
C     PROGRAM TELES.FOR
C     Makes a file 0000 that can be shown or printed
      WRITE(9,40)
40    FORMAT(/,'TELES    R,HM,F,PL',/)
      READ(9,*)R,HM,F,PL
      WRITE(6,41)R,HM,F,PL
41    FORMAT(5X,'TELES    R,HM,F,PL',//,4F10.2,//)
      DO 10 J=1,30
      H=FLOAT(J)*(HM/30.)
      TR=ASIN(H/R)
      SR=SIN(TR+TR)
      CR=COS(TR+TR)
      PB=H/SR
      A=PL-PB-R*COS(TR)
      E=(F-PB)/A
      PR=0.5*A*(1.-E*E)/(1.+E*CR)
      QX=PB-PR*CR
      QY=-PR*SR
      WRITE(6,42)H,QX,QY
42    FORMAT(3X,1F6.2,2F11.6)
10    CONTINUE
      END
```

```
C     PROGRAM RPLAN.FOR
      WRITE(9,40)
40    FORMAT(/,' RPLAN    R,HM,SM,FD,PL,XD ',/)
      READ(9,*)R,HM,SM,FD,PL,XD
      WRITE(6,40)
      WRITE(6,41)R,HM,SM,FD,PL,XD
41    FORMAT(2X,6F9.3,//)
      RYL=0.
      RXL=XD
      M=0
      DO 10 J=1,150
      M=M+1
      H=FLOAT(J)*ABS(HM/150.)
      W=ASIN(H/R)
      WS=SIN(W+W)
      WC=COS(W+W)
      WB=-H/WC
      X1=R*COS(W)
      Y1=H
      PR=PL-X1
      STH=FLOAT(J)*(SM/150.)
      THETA=ASIN(STH)
      STH=SIGN(1.,HM)*STH
      CTH=SIGN(1.,HM)*COS(THETA)
      X2=FD-PR*CTH
      Y2=PR*STH
      A=(X2-X1)/(Y1-Y2)
      B=0.5*(Y1+Y2-A*(X1+X2))
      X3=X1-PR*WC
      Y3=Y1-PR*WS
      X4=FD
      Y4=0.
      C=(X4-X3)/(Y3-Y4)
      D=0.5*(Y3+Y4-C*(X3+X4))
      PX=(D-B)/(A-C)
      PY=A*PX+B
      IF(J-1) 14,14,11
11    X5=0.5*(PXL+PX)
      Y5=0.5*(PYL+PY)
      A=(Y5-RYL)/(X5-RXL)
      B=RYL-A*RXL
      C=-STH/CTH
      D=-FD*C
      RX=(D-B)/(A-C)
```

```
      RY=C*RX+D
      RT=(RY-PY)/(RX-PX)
      T=(WS*STH-WC*CTH-1.)/(WC*STH+WS*CTH)
      A=(RT+T)/(1.-RT*T)
      B=PY-A*PX
      C=WS/WC
      D=Y1-C*X1
      SX=(D-B)/(A-C)
      SY=C*SX+D
      IF(M-5) 13,12,12
12    M=0
      WRITE(6,42)H,SX,SY,RX,RY
42    FORMAT(1F8.3,2X,2F11.6,2X,2F11.6)
13    RYL=RY
      RXL=RX
14    PXL=PX
10    PYL=PY
      END


C     PROGRAM MPLAN.FOR
      WRITE(9,40)
40    FORMAT(/,' MPLAN    AN,G,F,P,XD',/)
      WRITE(6,40)
      READ(9,*)AN,G,F,P,XD
      WRITE(6,41)AN,G,F,P,XD
41    FORMAT(2X,5F9.3,//)
      E=480./AN
      RYL=0.
      RXL=XD
      M=0
      DO 10 J=1,480
      M=M+1
      SA=FLOAT(J)/E
      SB=SA/G
      CA=SQRT(1.-SA*SA)
      CB=SQRT(1.-SB*SB)
      A=(F-P*CA)/(P*SA)
      B=0.5*(P*SA-A*(F+P*CA))
      C=(F-P*CB)/(P*SB)
      D=-0.5*(P*SB+C*(F-P*CB))
      PX=(D-B)/(A-C)
      PY=A*PX+B
      IF(J-1) 14,11,11
11    X1=0.5*(PXL+PX)
```

```
      Y1=0.5*(PYL+PY)
      A=(Y1–RYL)/(X1–RXL)
      B=RYL–A*RXL
      C=SB/CB
      D=–F*C
      RX=(D–B)/(A–C)
      RY=C*RX+D
      RT=(RY–PY)/(RX–PX)
      T=(SA*CB–CA*SB)/(1.+CA*CB+SA*SB)
      A=(RT+T)/(1.–RT*T)
      B=PY–A*PX
      C=SA/CA
      SX=B/(C–A)
      SY=C*SX
      ST=(SY–PY)/(SX–PX)
      IF(M–16) 13,12,12
12    M=0
      WRITE(6,42)SA,SX,SY,RX,RY
42    FORMAT(1F8.4,2X,2F11.6,2X,2F11.6)
13    RYL=RY
      RXL=RX
14    PXL=PX
10    PYL=PY
      END
```

# Appendix 2

Appendix 2 The program XROT.FOR is a Fortran code program for simulating image reconstruction for X-ray rotational aperture synthesis. The particular configuration is that of Figures 9 and 10 in Chapter 2.

The READ(,) and WRITE(,) instructions may have to be adapted to the user's hardware. On mine, unit 9 refers to the console and unit 6 refers to the disk.

In addition, the subroutines ERASE and DRAW(KX,KY,LX,LY) and the function IRAN(SEED) are hardware-dependent assembler-language routines that will have to be replaced to suit the user's hardware.

ERASE simply erases the screen.

DRAW(KX,KY,LX,LY) draws a straight line from point (KX,KY) to point (LX,LY) in a coordinate system having its origin at the upper left of the screen and (639,399) at the lower right. Note that the Y-scale goes downward.

IRAN(SEED) is a pseudo-random number generator giving uniformly distributed integers from 0 to 65535. IRAN(SEED) involves the CHARACTER SEED*8, which is 8 characters (64 bits) and so is not likely to be repeating within the life of the computer.

XROT first asks for the brightness and coordinates $B,X,Y$ of the first star in the source picture. The range for $X$ and $Y$ is $\pm\pi$, because a strong side lobe develops at a radius of $2\pi$. It then asks for $B,X,Y$ for subsequent stars until a star is entered with $B = 0$. The program next asks for the number $N$ of photons to be in the picture and the initialization of the 8-character SEED. If $N = 0$, the program synthesizes the image directly from the probability values as if there were an unlimited or infinite number of photons.

The height of the display is normalized, and a screen dump can be used to preserve the results, such as is seen in Figure 13 in Chapter 2.

```
C     PROGRAM XROT.FOR
      DIMENSION S(320),Z(1152),L(1152),M(1152),SS(256),P(128,32)
      CHARACTER SEED*8
      DATA ZS,PMAX,YP/0.,0.,0./
      PI=3.141593
```

```
      P2=PI+PI
      PH=0.5*PI
      PS=201.25*PI
      DO 10 J=1,80
10    S(J)=SIN(0.09817477*(J–1))
      CALL ERASE
12    WRITE(9,100)
100   FORMAT(' B,X,Y ')
      READ(9,*)B,X,Y
      IF(B.EQ.0.)GO TO 40
      DO 30 JT=1,64
      PR=X*S(JT+16)–Y*S(JT)
      DO 20 JR=1,9
      R=JR*PR+PS
      N=128*JR–128+JT
      Z(N)=Z(N)+B*ABS(PI–AMOD(R,P2))
      N=N+64
20    Z(N)=Z(N)+B*ABS(PI–AMOD(R+PH,P2))
30    CONTINUE
      GO TO 12
40    DO 42 J=1,1152
42    ZS=ZS+Z(J)
      ZS=–ZS/2304.
      WRITE(9,103)
103   FORMAT(' NT,SEED ')
      READ(9,*)NT,SEED
      IF(NT.EQ.0)GO TO 45
C     end of probability setup, start of Monte-Carlo
      Z(M)=0.
      DO 22 J=1,1152
22    ZM=MAX(ZM,Z(J))
      ZN=65536./ZM
      DO 23 J=1,1152
23    M(J)=INT(ZN*Z(J))
      ZS=–FLOAT(NT)/2304.
25    N=MOD(IRAN(SEED),1152)+1
      IF(IRAN(SEED).GT.M(N))GO TO 25
      L(N)=L(N)+1
      NT=NT–1
      IF(NT.GT.0)GO TO 25
      DO 43 J=1,1152
43    Z(J)=FLOAT(L(J))
C     start of Fourier synthesis
45    DO 41 J=1,320
41    S(J)=SIN(0.0245437*(J–1))
```

```
      CALL ERASE
      DO 60 JT=1,32
      DO 63 J=1,256
63    SS(J)=ZS
      DO 61 JR=1,9
      NA=128*JR-128+JT
      NB=NA+64
      NC=NA+32
      ND=NB+32
      HR=Z(NA)-Z(NB)+Z(NC)-Z(ND)
      HI=Z(NA)+Z(NB)-Z(NC)-Z(ND)
      JS=1
      DO 62 J=1,256
      SS(J)=SS(J)+HR*S(JS+64)+HI*S(JS)
      JS=JS+JR
62    IF(JS.GT.256) JS=JS-256
61    CONTINUE
      V=8.*S(4*JT+61)
      U=2.*S(4*JT-3)
      A=16.*V
      DO 70 JY=1,32
      B=A-64.*U
      DO 71 JX=1,128
      JB=INT(B)
      JB=MOD(JB+8192,256)+1
      P(JX,JY)=P(JX,JY)-SS(JB)
71    B=B+U
70    A=A-V
60    CONTINUE
C     end of Fourier synthesis, start of display
      DO 82 IX=1,128
      DO 81 IY=1,32
81    PMAX=MAX(PMAX,P(IX,IY))
82    CONTINUE
      PN=150./PMAX
      DO 92 J=1,128
92    S(J)=0.
      DO 93 IY=1,32
      KX=64
      MY=360
      KY=MY-6*IY
      DO 94 IX=1,128
      S(IX)=MAX(S(IX),PN*P(IX,IY)+6.*IY)
      LY=MY-INT(S(IX))
      LX=KX+4
```

```
      CALL DRAW(KX,KY,LX,LY)
      KX=LX
94    KY=LY
93    CONTINUE
      PAUSE
      END
```

# Index